Mass and Energy Balancing

Mass and Energy Balancing
Calculations for Plant Design

David Pritchard and Shaik Feroz

CRC Press
Taylor & Francis Group
Boca Raton London New York

CRC Press is an imprint of the
Taylor & Francis Group, an **informa** business

MATLAB® is a trademark of The MathWorks, Inc. and is used with permission. The MathWorks does not warrant the accuracy of the text or exercises in this book. This book's use or discussion of MATLAB® software or related products does not constitute endorsement or sponsorship by The MathWorks of a particular pedagogical approach or particular use of the MATLAB® software.

First edition published 2021
by CRC Press
6000 Broken Sound Parkway NW, Suite 300, Boca Raton, FL 33487-2742

and by CRC Press
2 Park Square, Milton Park, Abingdon, Oxon, OX14 4RN

Library of Congress Cataloging-in-Publication Data
Names: Pritchard, David (Chemical engineer), author. | Feroz, Shaik, author.
Title: Mass and energy balancing : calculations for plant design / authored by David Pritchard and Shaik Feroz.
Description: First edition. | Boca Raton, FL : CRC Press, 2021. | Includes bibliographical references and index. | Summary: "The text provides a comprehensive set of calculations relating to mass and energy balances for an entire process plant. An ammonia synthesis plant is taken as a calculation model to develop the relevant mass and energy balances necessary for the design and subsequent production. Instead of teaching the basics, the text gives a detailed series of process integrated and illustrated calculations to help readers develop and design a process plant. This book will serve undergraduate Chemical Engineering students as a teaching aid in capstone design and related courses and gives useful insights to advanced students, researchers, and industry personnel within the Chemical Engineering field"— Provided by publisher.
Identifiers: LCCN 2020055140 (print) | LCCN 2020055141 (ebook) |
ISBN 9780367709679 (pbk) | ISBN 9780367710798 (hbk) | ISBN 9781003149200 (ebk)
Subjects: LCSH: Chemical process control. | Chemical plants—Design and construction—Mathematics. | Heat transfer. | Mass transfer. | Engineering mathematics—Formulae.
Classification: LCC TP155.75.P736 2021 (print) | LCC TP155.75 (ebook) | DDC 660/.2815—dc23
LC record available at https://lccn.loc.gov/2020055140
LC ebook record available at https://lccn.loc.gov/2020055141

ISBN: 978-0-367-71079-8 (hbk)
ISBN: 978-0-367-70967-9 (pbk)
ISBN: 978-1-003-14920-0 (ebk)

Typeset in Times
by codeMantra

Contents

Foreword

I am delighted to find this book is not only another book on mass and energy balances, but a book written by keeping in mind the needs of both the students and the industrial design engineers.

This book provides a clear understanding of mass and energy balances of a process plant. The stepwise procedure required for performing the mass and energy balance calculations for various process units has been considered and explained in an easy way. All the academic and industrial experience of the authors seems to be distilled out in this book. This book is intended as a college- or university-level text for students in chemical engineering and related fields. It is also detailed enough in its development of each section to be useful as a reference for practising plant design engineers.

I am sure this book will go a long way in providing the requisite knowledge for chemical and process engineers. I am also confident that it will be received and well read by the professionals and students all over the world. I wish the book and the authors all success.

– Prof. Detlef Bahnemann

Foreword

Preface

The purpose of this book is to provide an example of detailed mass and energy balances carried out on an existing process. This sort of calculation would, typically, be required as part of design calculations. Many books that are used to teach mass and energy balancing illustrate the various pieces of knowledge required through a series of relevant, but not necessarily integrated, calculations. This book will not claim to do this, and the various pieces of knowledge required will be assumed to be in place. What the text will do is give a detailed series of process integrated and illustrated calculations. It will attempt to guide the reader as to how the knowledge that they should have can be realistically applied, and the series of integrated calculations will lead to the production of a specified amount of final product.

Some applications of prior assumed knowledge will be laid out in 'teaching' appendices. These will be able to be used in conjunction with the main text where the figures calculated may need a more detailed explanation. This book would, typically, be used by first- and second-year students as a teaching aid but could give some useful insights to other students, researchers and industry personnel.

In considering the text, the reader should be aware that they have a responsibility. Any attempt to demonstrate mass and energy balances means that it is important that the numbers are understood. That might appear obvious, but it is easy just to see figures without realising where they came from; an attempt should at least be made to try to check the genesis of the figures.

Any contact with the energy balance in the text will immediately indicate that the enthalpy values to be used are usually calculable using a polynomial giving component enthalpies as a function of temperature. Finding the enthalpy value can then become almost a mechanical exercise. How exactly enthalpies are calculated for energy balances differs according to sources of data. The important thing in this text is that sources of enthalpy data are discussed and basic equations explored. If these sections are properly checked, then the background given usually makes the calculation of enthalpies more than just a mechanical act. At least, unlike some mass and energy balances, some data are available.

Usually, the data for enthalpy are given in kJ/mol for component enthalpies and kJ for stream enthalpies. Where appropriate, the mole is sometimes used, but usually the text deals in kmol and kJ.

It is always interesting to follow the latest convention. The text had to deal with desulphurisation, desulfurisation or desulfurization. This text has adopted desulphurisation.

A short examination of the book text will reveal a range of tables with a variety of numbers in each table. Given the fact that mass and energy balances for a range of operations are being calculated, this is inevitable. These values are presented in different formats, but in this text, numbers are in the so-called scientific format where numbers are expressed in exponential form. The number of significant figures is a matter for debate. It is acknowledged that the number of significant figures should not show the excesses of spreadsheets sometime seen. Large numbers are rounded off in scientific notation.

In this text, the number of significant figures in energy balances is set at 5. This is illustrative and a compromise for the data being used and the data that may be used as it is available. In this text, various data sources for enthalpy are quoted and the figures used in the main enthalpy polynomials are reproduced. Using the basis of the mass balance quoted, relatively large numbers can be generated. The number of significant figures can be decided by the nature of the data to be used and processed. The basis in this text is fixed and is not the main purpose of the text. The balances to be carried out are illustrative and can be adapted as necessary.

A text that attempts to carry out multiple calculations can be difficult to follow. It is impossible to illustrate every step taken in a calculation, so there has to be an element of self-instruction and individual responsibility in tracing from where and why particular numbers are produced. This is particularly true in a text of the type shown here. The appendices contain material that should be studied in conjunction with the calculations being carried out.

The production of the syngas can be followed on the block diagram supplied. It is vital that this is available as the mass and energy balances progress. The various chapter contents and process progression can be followed using the block diagram, and it is vital that this is used as the guide as progress is made through the chapters.

It will be observed that Chapter 5 contains fairly standard, orthodox questions that do relate in many instances to the ammonia synthesis loop. There are usually no 'extra' principles involved in solving these although the manner in which they are laid out tries to be supportive in producing answers. In teaching, I always found that many learners were adept at fitting the answer to the question, and it was a mistake to supply too many details. That has not happened here, and there is a challenge as to how an answers chapter should be used.

In many ways, this is an experimental text with possible attendant problems. As has been pointed out, it is worth taking time to go through the calculations carried out and try to follow the theoretical logic of what is laid out. There is enough information supplied in the text and appendices for you to attempt to carry out the calculation. This challenges you to reproduce your own version of the answer and not necessarily to reproduce the answer in the text.

You should find boxes attached to each chapter which initially try to identify what will be calculated in the chapter. The box at the end of the chapter tries to summarise the parameters that have actually been calculated. This is a good exercise in deciding if you understood how they were obtained and in deciding which other parameters you might wish to be included or even omitted.

Good luck on your journey.

MATLAB® is a registered trademark of The MathWorks, Inc. For product information, Please contact:

The MathWorks, Inc.
3 Apple Hill Drive
Natick, MA 01760-2098 USA
Tel: 508-647-7000
Fax: 508-647-7001
E-mail: info@mathworks.com
Web: www.mathworks.com

Acknowledgements

This text could not have come into existence without the preliminary work carried out by Graham Hampson at Teesside University in the United Kingdom. All his work is acknowledged here.

Gabrielle Vernachio and Allison Shatkin of CRC have offered consistent support, advice and action.

One of us (David Pritchard) expresses thanks to his wife, Barbara, for her overheated calculator and its use and the many hours spent in checking text and producing an index. Intellectual challenge comes in many ways and guises.

To many friends and colleagues over the years who have discussed students, their problems and learning needs and possible solutions, due acknowledgement is made.

Acknowledgements

Nomenclature

a, b, c, d	Polynomial coefficients
a, b, c, d	Stoichiometric coefficients
a_i	Activity of component i
C_P	Specific heat capacity at constant pressure
C_V	Specific heat capacity at constant volume
f_i	Component fugacity
f_i^θ	Standard component fugacity
$\left(\Delta G_T\right)_f$	Gibbs free energy change of formation at temperature T
ΔG_T	Gibbs free energy change at temperature T
ΔG_T^θ	Standard Gibbs free energy change at temperature T
H	Power requirement
ΔH	Enthalpy change
ΔH_F^0	Standard enthalpy of formation
H_0^o	Standard enthalpy related to a temperature of $0°K$.
K_P	Equilibrium constant
M	component moles
N	Number of components
n	Polytropic index
n	Number of moles
P	Pressure
p_i	Partial pressure of component i
$P*$	Standard pressure
q	Energy transfer as heat
T	Temperature
ΔU	Internal energy change
w	Energy transfer as work
y_i	Gas-/vapour-phase mole fraction of component i
z_i	Fractional conversion of component i
$\gamma = C_P/C_V$	

Authors

Dr. David Pritchard is a retired academic who continues to work in academic areas of chemical engineering. He obtained his first degree from the Manchester College of Science and Technology, Faculty of Technology in the University of Manchester, UK. He obtained his doctorate from Bath University of Technology (now University of Bath), UK. On graduation, he initially worked in the process industries for British Titan (subsequently Tioxide International, UK) as a research officer looking to improve the plant engineering. From here, he joined the then Teesside Polytechnic, Department of Chemical Engineering, UK, as a lecturer. The Polytechnic was subsequently incorporated as the University of Teesside and still exists as such.

At Teesside, he held every undergraduate course leadership post that then existed (not all at the same time) and was responsible for course development and, with colleagues, produced a pioneering first-year course on the chemical engineering degree. This first year obtained awards for the novelty of the teaching approach including recognition from the RSA. He lectured chemical engineering at every level from undergraduate diploma (HND, HNC), degree (BSc, BEng), postgraduate (MSc) and design projects. He helped pioneer the introduction of part-time routes for degrees in chemical engineering and managed many students studying on ERASMUS schemes in receiving degrees under a special, externally examined, pathway.

He also carried out research into a range of areas particularly in the fields of separation processes and phase equilibria. He produced over 30 refereed research papers together with a number of conference papers and supervised over 20 PhD students and a number of MSc students. Within the newly formed School of Science and Technology, he took responsibility for the running of the Division of Chemical Engineering and oversaw various academic initiatives. He acted as an external examiner for various UK courses at undergraduate and postgraduate levels and was a PhD external examiner.

Dr. Shaik Feroz is currently working as Research Professor at Prince Mohammed Bin Fahd University, Kingdom of Saudi Arabia. Dr. Feroz obtained his doctorate in the field of chemical engineering from Andhra University, India, in 2004; Postdoc Research Fellowship from Leibniz University, Germany, in 2015; M.Tech in chemical engineering from Osmania University, India, in 1998; B.Tech in chemical engineering from S. V. University, India, in 1992; and Post-Graduate Diploma in Environmental Studies from Andhra University in 2003.

Dr. Feroz has expertise in process engineering, plant design and troubleshooting, quality control using advance analytical equipment, wastewater treatment, solar energy systems (PV and CSP) for desalination, hot water systems and water treatment, synthesis of nano-photocatalysts, simultaneous treatment of wastewater and production of hydrogen, environmental impact assessment, and design, delivery and management of chemical and process engineering programmes and tailor-made industrial-based training programmes related to chemical and process engineering.

Dr. Feroz has more than 170 publications to his credit in journals and conferences of international repute and supervised four PhD research works with another three ongoing. Dr. Feroz is associated as the Principal Investigator/Co-Investigator to a number of research projects and also involved as "Technology transfer agent" by Innovation Research Center, Sultanate of Oman. He has a significant contribution as editorial member and peer reviewer for various reputed international journals and conferences. He was awarded "Best Researcher" by the Caledonian College of Engineering for the year 2014. Dr. Feroz has total experience of 27 years in teaching, research and industry.

1 Introduction, Reformers and Stream Energy Interchange

1.1 GENERAL INTRODUCTION AND BASIS FOR CALCULATION

If you are intending to use this text in an informative, learning, teaching or critical situation, then there is an assumption that you know some of the principles of process engineering and the breadth and depth of the challenge you may be faced with. There exist a number of texts [1] that can supplement your knowledge and be informative in introducing you to further aspects of process engineering. This particular text deals with one of the initial, essential, procedures in process engineering, an analysis of the mass and energy requirement of any and all operations.

You cannot carry out mass and energy balances unless the basic principles of thermodynamics are understood and observed. The first law is often pithily stated that mass is conserved. There are obviously various subheadings to write here, but the conservation of mass sounds a good basis for

$$\text{Input} = \text{Output}$$

Those of you who have experienced process engineering in theory and in practice will know that it is crucial to be able to handle the masses and energies involved in the chemical reaction as well as non-reacting systems. As usual, thermodynamics takes us calmly by the shoulder and points out that species may change through the reaction, but the total mass of atoms confirms that mass and energy input does indeed equal mass and energy output. The second law of thermodynamics gives us the concepts and tools to be able to carry out quantitative analysis of the balance when chemical reaction is also involved.

Many operations within process engineering are carried out at steady state, and the mass and energy balances do not vary with respect to time; for both mass and energy, the input equals the output. Mass and energy balances have a time-variant form if required; a time-variant term, the accumulation or depletion, can be added or subtracted. This term will not be considered in this text. All production of ammonia syngas is considered at steady state.

If you are relatively new to process engineering and its component parts, then it might be useful to read the next few pages. If you know everything there is to know about it, then the next few pages may be old hat – but still readable. A general description of process engineering would include the following, this description includes concepts already covered:

Process engineering is the engineering part of production processes in which chemical and physical changes take place. The process engineer deals with all aspects of the chemical industry from planning, designing, construction, operation and control, as well as research and development, marketing and technical services for customers. Of course, the process engineer should also be aware of the economic aspects of production. Process engineering is aimed towards the design of processes that change materials from one form or species to another, more useful and valuable form, economically, safely and in an environmentally acceptable way.

All engineers usually employ mathematics, physics and the 'engineering art' to overcome technical problems in a safe and economical fashion; usually, it is the process engineer alone that draws upon the powerful science of chemistry to solve a wide range of problems.

In practice, process/chemical engineering is the application of basic sciences (maths, chemistry, physics and biology) and engineering principles to the development, design, operation and maintenance of processes to convert raw materials to useful products and, at the same time, improve the human environment. Process/chemical engineering involves specifying equipment, operating conditions, instrumentation and process control for all these changes. Process/chemical engineers translate processes developed in the lab into practical applications for the production of products such as plastics, medicines, detergents and fuels; design plants to maximise productivity and minimise costs; and also evaluate plant operations for performance and product quality.

Thus, as stated, the knowledge base for process/chemical engineers starts from a basic understanding of chemical engineering principles with a strong chemistry back ground; material and energy balance calculation, thermodynamics principles, momentum, heat and mass transfer principles, mechanical separation principles, manufacturing process flow diagrams; reaction engineering principles, process instrumentation and control principles, process equipment and plant design principles, and health and safety management principles; and management and economics. In addition to this, the knowledge base is extended to complementary areas such as modelling and simulation principles, optimisation principles, environmental engineering principles, corrosion engineering principles, and nanotechnology.

A structure of the logic used for synthesising and analysing processing schemes in the chemical and allied industries is essential. Thus, the basic concept adopted is that all processing schemes can be built into a series of individual, or unit, steps. Usually, if a step involves a chemical change, it is called a unit process; if physical change, a unit operation. The basic tools of the chemical engineer for the design, study or improvement of a unit process are the mass balance, the energy balance, kinetic rate of reaction and position of equilibrium (the last needs to be included only if the reaction does not go to completion).

In manufacturing, a unit process is a single component part of the end-to-end manufacturing process that transforms raw materials into finished goods. Unit processing is the basic processing in process/chemical engineering. Together with unit operations, it forms the main principle of the varied chemical industries. Each genre of unit processing follows the same chemical law much as each genre of unit operations follows the same physical law.

Material balances are one of the most basic and useful tools in the process/chemical engineering field. The principles involved have been set out previously in this introduction. One of the first things that chemical engineers need to know about processes is the many different ways in which a process may be operated. There are three major classifications of processes:

i. In a batch process, material is placed in the vessel at the start and (only) removed at the end – no material is exchanged with the surroundings during the process. Examples are baking bread, fermentations and small-scale chemical production (pharmaceuticals).
ii. In a continuous process, material flows into and out of the process during the entire duration. Examples are pool filter and distillation processes.
iii. A semi-batch process is one that does not neatly fit into either of the other categories and often contains elements of both (i.e. it is a catch-all classification). Examples are a washing machine and a fermentation with purge.

Again, as previously indicated, each of the above classes of process may be further distinguished by how they operate with respect to time. For a steady state, none of the process variables change with time (if we ignore small, random fluctuations); conversely, a process may be run at unsteady state if the process variables change with time.

Energy balances are used to quantify the energy used or produced by a system. Energy balances can be calculated on the basis of external energy used per kilogram of product, or raw material processed, or on dry solids or some key component; a basis must be stated in carrying out the balance.

The first law of thermodynamics enables us to understand the mechanisms involved in energy transfer (heat and work transfer) between a system and its surroundings and produces thermodynamic properties and functions that enable us to quantify these transfers. At constant pressure, the thermodynamic property *enthalpy* is defined and used. This property, at constant pressure, can be shown to quantify the energy transfer as heat and hence is used in the energy balance. Enthalpy is conserved and, as with the mass balance, enthalpy balances can be written round the various items of equipment in the process, either in process stages or round the whole plant; it is assumed that no appreciable heat is converted to other forms of energy such as work. Enthalpy (H) is always referred to some reference level or datum, so that the quantities are relative to this datum. Working out energy balances is then a matter of considering the various quantities of material involved, their specific heats and their changes in temperature or state (as quite frequently energies of vaporisation (latent heats) arising from phase changes are encountered). In the problems you will deal with when ammonia syngas production mass and energy balances are considered, enthalpy values are required, and the availability of the necessary data is discussed in Appendix 3.

In basic consideration of the process engineering, it is important to be able to deal with the mass and energy balances in terms of how far a chemical reaction will go under fixed conditions. Again thermodynamics, through the Gibbs energy and the equilibrium constant, can help to fix the final balances in the plant.

For other specific elements of the plant, thermodynamics can be used to do meaningful calculations on various unit operations; thermodynamics can provide the equilibrium and enthalpy data needed for design of distillation columns, absorption columns, evaporators, condensers and other units where heat exchange is involved.

Many chemical processes are concerned with the problem of changing composition of solutions and mixtures which may or may not involve chemical reactions. These operations are directed towards separating a substance into its components, e.g. filtration and screening, which are entirely mechanical operations.

1.2   INTRODUCTION TO THE MASS AND ENERGY BALANCES FOR THE PRODUCTION OF AMMONIA SYNGAS

Unless otherwise stated, the enthalpies used in the *energy balances* are calculated from polynomial expressions. In Appendix 3, you will find some basic ideas explained and will have access to the polynomial expression constants used in calculations. Now may be the time to look at that appendix and get the basis of the enthalpy property and the various ways it can be calculated.

This text now attempts to set out some basic mass and energy balances related to the production of *ammonia synthesis gas (syngas)*. Obviously, this then relates to the production of ammonia! This text does not attempt to give the history, usefulness and detailed listings of the properties and uses of ammonia. We start at the basic point that we require to produce a synthesis gas that can subsequently be used to produce ammonia gas, and we are aware that we need to consider the basic stoichiometric equation:

$$N_{2(g)} + 3H_{2(g)} \leftrightarrow 2NH_{3(g)}$$

The production of ammonia essentially breaks down into three sections:

1. Synthesis gas production
   a. Feedstock pre-treatment and gas generation.
   b. Carbon monoxide conversion.
   c. Gas purification.
2. Compression
3. Synthesis

In *synthesis gas production*, the aim is to produce a pure mixture of nitrogen and hydrogen in the stoichiometric ratio 1:3 from an appropriate feedstock.

This text is based on the production of the synthesis gas using a process based on the block diagram shown below (Figure 1.1):

It is worth noting that the basic units making up the block diagram are the items on which we will be carrying out detailed mass and energy balances.

Basically, Step (a) as described above is set out as a series of operations in the block diagram.

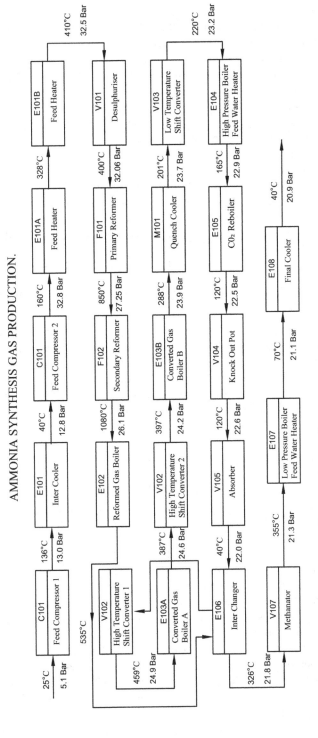

FIGURE 1.1 Block diagram for the production of ammonia syngas.

In the diagram, the following convention is followed:

Items marked E are involved with heat exchange.
Items marked C are involved with compression.
Items marked F are furnaces.
Items marked V are process vessels.

If we check the sections set out for ammonia production, i.e.:

a. Feedstock pre-treatment and gas generation.
b. Carbon monoxide conversion.
c. Gas purification

then on the block diagram, Item C101 (where the primary feedstock, natural gas, enters the process via a compressor) to Item E102 (where energy from the initial production of the N_2/H_2 mixture is exchanged to provide energy elsewhere in the process) cover 'Feedstock pre-treatment and gas generation'.

Item V102 (where CO is removed via the water–gas shift reaction)

$$CO + H_2O \leftrightarrow CO_2 + H_2$$

to Item E104 (where energy from the water–gas shift converters is used to input energy to help to raise steam) relate to the carbon monoxide conversion.

Item E105 (where energy is used to recover solvent for absorber V105 by removing CO_2) to Item V105 (where CO_2 is actually removed in an appropriate solvent) cover most of the CO_2 removal. Item V107, the methanator, removes vestiges of CO and CO_2 left in the gas stream by converting to methane through the reactions:

$$CO + 3H_2 \leftrightarrow CH_4 + H_2O$$

$$CO_2 + 4H_2 \leftrightarrow CH_4 + 2H_2O$$

With some more heat exchange, the syngas is then ready for the ammonia synthesis loop.

The scene is set, and the chapters and sections that follow deal with this mass and energy balancing in the production of ammonia syngas. Each section deals with one of the sequential elements shown in the block diagram for the process. Each section takes the reader through the balances required for the mass and energy in the particular operation and tries to point out some of the other factors that can operate when carrying out the mass and energy balances.

The mass and energy balances can be carried out but, occasionally, it is helpful to describe the operation being carried out in slightly more detail. No attempt is made to show the detailed process design involved in the operation. Again, it is emphasised that the text assumes that the basic elements of mass and energy balancing are familiar to the reader.

In carrying out the energy balancing, a reference has been made to Appendix 3 in which elements of the approaches to calculation of enthalpy and enthalpy differences are set out. In some ways, the calculation of enthalpy and enthalpy differences has been 'simplified', by use of polynomials for component enthalpies. The basis of the polynomials is set out and the basis of temperature explained. The calculation of the enthalpy change from the defining equation:

$$\Delta H = \int_{T_1}^{T_2} mC_P \, dT$$

is illustrated by its use in the section dealing with the desulphurisation section of the process.

Because each chapter with its sections is in itself an exercise in carrying out a mass and energy balance, no other tutorial exercises are set in the section. Each section schedules the calculations required and how it is proposed to carry them out.

At the beginning of each section, a box laying out the basic calculational requirements of the section is presented. At the end of each section, a box containing the figures calculated is presented and a brief discussion of their significance given.

In some of the other chapter sections, there are other calculations presented; e.g., is the water vapour in the gas stream likely to condense at the pressure and temperature figures given? This usually requires some estimate of a gas/vapour dew point. There is a presupposition that basic ideas related to Dalton's law and partial pressures are known and even some understanding of dew point is assumed. There is a brief explanation of Dalton's law in Appendix 1. Knowing some basic principles, dew point calculations are presented as appropriate.

Because the main emphasis of the text is calculation, it is necessary to carry out the numerical calculations as appropriate. In today's calculational world, there is a range of hardware/software available. A perusal of the text will show that things have been kept basic and relatively straightforward. The calculations have normally been carried out using Microsoft Excel spreadsheets. The necessary equations are usually made clear in the text, e.g. Appendix 2. It is made quite clear that any appropriate calculational aids can be used to carry out calculations and check numerical answers.

1.3 THE BLOCK DIAGRAM FOR PRODUCTION OF AMMONIA SYNTHESIS GAS

As the syngas stream progresses through the processing, the major mass and energy balancing operations are considered to be the following:

1. Desulphuriser (Item V101)
2. Primary reformer (Item F101)
3. Secondary reformer (Item F102)
4. HTS converter (Item V102)
5. LTS converter (Item V103)
6. Absorber (Item V105)
7. Methanator (Item V107)

Most of the operations will require an element of mass balancing, but the bulk of those scheduled in this list involves mass balancing with chemical reaction. The absorber involves internal streams inside the mass balance envelope, and this will need individual and overall mass balancing.

For energy balance, most of the operations within the block diagram have energy transfers involved. In addition to the major mass and energy balancing operations already identified, there are other operations that require balances to be carried out. The raising of steam using the energy from the syngas stream and the use of the syngas stream energy for heating purposes are major operations requiring energy balances.

The main operations shown on the block diagram involving further energy transfers are scheduled below:

1. Intercooler (E101)
2. Feed heater (E101A)
3. Reformed gas boiler (E102)
4. Converted gas boilers A and B (E103A,B)
5. Quench cooler (M101)
6. High-pressure boiler feed water (BFW) heater (E104)
7. CO_2 reboiler (E105)
8. Interchanger (E106)
9. Low-pressure BFW heater (E107)
10. Final cooler (E108)

As has been stated, the calculations of the required mass and energy balances will be laid out in sections that follow the block diagram.

1.4 SIMPLE PRINCIPLES OF MASS BALANCING

If you work through a text on mass balances, you should be introduced to some basics, particularly the definition and calculation of the mole. It is assumed you can define the mole, calculate it and use it. In using it, you should be aware of its significance in stoichiometric equations used in representing and balancing equations for chemical reactions.

Essentially, we will do all our calculations in the text at steady state, and this leaves us with the easy statement of conservation of mass previously cited:

$$\text{Mass Input} = \text{Mass Output}$$

Obviously, such a statement relates to the mass of each species entering and leaving. Thus, the simplest statement is to recognise that all atoms of an element entering a defined system must balance the number of atoms of the same element leaving the system. This balance can be written for all elements entering and leaving the system. We also recognise that if, for a particular species, e.g. a molecule, we express its mass in moles, then we know that the moles have been calculated relative to the species molar mass; thus, whilst in the input/output balance the masses will balance, the moles do not necessarily balance. If we express the moles as masses in appropriate mass units, we have to find that input masses equal output masses. On this principle, the stoichiometric reaction equation is a simple mass balance statement.

We should know that in mass balancing, where we have to consider chemical reaction, as we work through the balance, the operation can also include recycling unreacted components usually in the chemical reactor. To many meeting this operation, it appears preposterous that you can return material to the input of the reactor without getting some sort of build-up in the reactor. As long as it is understood that the feed entering the reacting system before receiving a recycle stream must balance the product leaving the reacting system through the stoichiometric equation and that input and output points for the recycle can be balanced at the various exit and entry points through a junction balance, recycles can be mass-balanced with relative ease, and the presence of an accumulation or depletion contradicting the steady state can be dismissed.

In the production of ammonia syngas, as we experience it, no recycle calculations are required, but it is vital that atom and component balances are understood and can be applied. This short section will be referred to within the mass balances that are carried out in the chapter sections below. In doing the mass balances, various data are required, and these will be reproduced and referenced appropriately. You may know by now that in carrying out mass balances in industrial problems, the calculations are often as good as the data available.

1.5 DESIGN BASIS

The simplest design method is to use a basis of 100 kmol feedstock and work through the process step by step to determine the amount of product. For a specified flow rate of nitrogen/hydrogen product, everything can be scaled accordingly. In these calculations, everything is related to 100 kmol of feedstock. In addressing the energy balance, the first problem that will arise is specification of the feedstock compressor and pumps since the efficiency depends on the flow rate.

Estimates of pressure drop will not separate pressure drops in equipment from pressure drop in pipework. At the level that these calculations are presented, this would not be practical.

Part of the process requires the removal of carbon dioxide after the water–gas shift conversion. This chapter section will not be considered in detail. It will be assumed that the process operates with CO_2 removal by absorption. It is recognised that other methods are available, e.g. pressure swing adsorption, but for mass balancing purposes, we will use the absorption technique. This can depend on licensor's data, and assumptions will be made. The solution from the absorption may well be circulated by pumps driven by steam turbines; these are ignored in the steam system. It will be implicitly assumed that the pumps are electrically driven.

Heat losses will be allowed in the energy balances, usually on a sensible, arbitrary basis, since neither the actual capacity of the plant nor the detailed equipment design is known in the calculations laid out.

As said, the assumption is that the feedstock is 100 kmol of a natural gas. The compositions of such a gas can vary considerably, but the assumption in this work will be made that the source is still similar to North Sea gas and the composition taken is broadly representative of this feedstock (Table 1.1):

TABLE 1.1

Composition of Undesulphurised and Desulphurised Natural Gas

Undesulphurised	Component	%v/v
	CH_4	94.68
	C_2H_4	3.00
	C_3H_8	0.50
	C_4H_{10}	0.40
	N_2	1.40
	H_2S	0.02
Desulphurised	**Component**	**%v/v**
	CH_4	94.70
	C_2H_4	3.00
	C_3H_8	0.50
	C_4H_{10}	0.40
	N_2	1.40

Supply Conditions

Pressure = 5.1 bar
Temperature = 25°C

1.6 PRIMARY REFORMER (F101 ON THE BLOCK DIAGRAM)

Basically, the natural gas is first desulphurised, and it will be seen from the block diagram that there are a number of operations in place to remove sulphur from the entering natural gas.

Once the principles of mass and energy balancing adopted in this text have been demonstrated on the primary reformer, these principles will be applied to the rest of the block diagram up to the final cooler.

There is a saying in English about putting the cart before the horse and that may be applied to the consideration of compression and desulphurisation of the natural gas feedstock after the rest of the syngas production process has been considered. This has been done because the approach being adopted in the text, particularly in the energy balance, was better illustrated by the approach adopted and calculations carried out on the primary reformer and subsequent operations.

The compression and desulphurisation chapter following the final cooler operation in the syngas production carries out appropriate mass and energy balances, includes a brief description of the compression and does energy balances illustrating a number of approaches referenced in Appendix 3.

REQUIRED

Component mass balancing.
Understanding and calculation of equilibrium constants.
Calculation of partial pressures using Dalton's law.
Definition and manipulation of reactant fractional conversions.
Setting conditions for endo- and exothermic reactions.
Calculation of fuel and energy requirements for an overall endothermic reactor.
Solving for fractional conversions by solving simultaneous, non-linear equations.

A nickel catalyst is employed for the reforming, and under the conditions described, the water–gas shift reaction is considered to reach thermodynamic equilibrium. For the methane–steam reaction, in doing this mass balance, an approach to equilibrium is used (see Appendix 5). In this case, for the conditions and catalyst used, a 30°C approach to equilibrium is assumed. Such information will be used in calculating on the basis of known thermodynamic data.

The approach to equilibrium is an attempt to quantify the performance of the catalyst being used in a reactor. The actual exit product compositions can be used to calculate an equilibrium constant, and hence, using appropriate $\ln K_P$ vs $1/T$ data, an equivalent value of T, the temperature, can be calculated. The difference between this value and the actual product temperature assumed is called the approach to equilibrium. Thus, for the methane–steam reaction, we will assume an approach to equilibrium of 30°C in an attempt to represent a realistic figure related to the catalyst used. The actual operation will be illustrated in the mass balance on the primary reformer.

The appropriate data are, at least, the equilibrium constants for Reactions 1 and 2, and these are set out in *Appendix 1* and their genesis given.

For the gaseous reactions discussed, we could write:

$$K_{P_1} = \frac{p_{CO}\ p_{H_2}^3}{p_{CH_4}\ p_{H_2O}}$$

$$K_{P_2} = \frac{p_{CO_2}\ p_{H_2}}{p_{CO}\ p_{H_2O}}$$

The equilibrium constant, as we should know, arises directly from the application of the second law of thermodynamics to the achieving of equilibrium for the chemical reaction. In this text, we write the equilibrium constant, K_P, in terms of partial pressures. We are assuming ideality in the gaseous phase. The partial pressures, p_i, are easily understood as the proportion of the total pressure exerted by any component i. The proportionality is expressed through the value of y_i, i.e. $p_i = y_i P$, where y_i is a component i mole fraction in the gaseous phase and P is the total system pressure.

In the published data on the K values for these reactions, there is usually agreement on the K values for the water–gas shift reaction, but there is some variation on the K values for the methane–steam reaction.

The values used in this work are calculated from the equations given below that give the K values as a function of absolute temperature T. Data from other sources quoted in Appendix 1 have similar $\ln K_p$ vs $1/T$ equations fitted; none of the equations fitted show a particular advantage over the others, so the stated equations, which had been used for many of the calculations, were employed.

1.6.1 METHANE–STEAM REACTION

$$\ln K_{P_1} = 30.42 - \left(\frac{27,106}{T}\right)$$

1.6.2 WATER–GAS SHIFT REACTION

$$\ln K_{P_2} = -3.798 + \left(\frac{4160}{T}\right)$$

There is a problem in steam reforming that, under the conditions of the reforming, carbon can also be formed by various reactions; this carbon formation can lead to deposition on the catalyst surface and reaction is inhibited at the surface. These reactions are usually supressed by increasing the steam/methane ratio above the stoichiometric ratio of 1. An inspection of the stoichiometric equation for the methane–steam reaction should lead to an appreciation that excess steam, i.e. steam above the required stoichiometric amount, will act as a diluent and increase the total moles at equilibrium. In terms of the calculation of reactant and product moles at equilibrium, this should increase the conversion of methane in this reaction. In the calculations outlined here, the steam/carbon ratio is taken as 3.5/1. In many modern plants operating under different conditions, this can be operated at a lower ratio.

In this text, we will use a fractional conversion, z, to quantify the amount of a given reactant that has been converted. Thus, the fractional conversion will be:

$$z_{\text{reactant}} = \frac{\text{Initial moles of reactant} - \text{moles of reactant at the end of reaction}}{\text{Initial moles of reactant}}$$

We define a methane conversion in Reaction 1 as z_{CH4} and a carbon monoxide conversion in Reaction 2 as z_{CO}; then, we can write the following table (Table 1.2) using 100 kmol of the originally tabled natural gas as the feedstock. To simplify the approach, the higher hydrocarbons are accounted as methane. In quoting the z values in Table 1.3, it is necessary to check that the definition of the fractional conversion is applied to the figures quoted. Being independent of path, it is assumed that the reactions can be considered in series, i.e. Reaction 1 followed by Reaction 2. This produces the moles quoted in Table 1.3:

A simple atom balance on the feedstock gives the following:

TABLE 1.2
Atom Balances on Some Natural Gas Components

Component	%v/v	C	H
CH_4	94.70	94.7	$94.7 \times 4 = 378.8$
C_2H_6	3.00	6.0	$3 \times 6 = 18$
C_3H_8	0.50	1.5	$0.5 \times 8 = 4$
C_4H_{10}	0.40	1.6	$0.4 \times 10 = 4$
N_2	1.40	0.0	0
Total		103.8	404.8

TABLE 1.3
Primary Reformer Analysis using Component Fractional Conversions

Component	Initial Moles (kmol)	Moles after Reaction 1 (kmol)	Moles after Reaction 2 (kmol)
CH_4	94.68	$98.6 - 98.6z_{CH4}$	$98.6 - 98.6z_{CH4}$
C_2H_4	3.00		
C_3H_8	0.50		
C_4H_{10}	0.40		
N_2	1.40	1.4	1.4
CO		$98.6z_{CH4}$	$98.6z_{CH4} - 98.6z_{CH4}z_{CO}$
H_2O	363.3	$363.3 - 98.6z_{CH4}$	$363.3 - 98.6z_{CH4} - 98.6z_{CH4}z_{CO}$
CO_2			$98.6z_{CH4}z_{CO}$
H_2		$3 \times 98.6z_{CH4}$	$3 \times 98.6z_{CH4} + 98.6z_{CH4}z_{CO}$
TOTAL			$463.3 + 2 \times 98.6z_{CH4}$

Thus, for a steam/carbon ratio of 3.5, we will take the steam flow rate as $103.8 \times 3.5 = 363.3$ kmol.

At first sight, this table looks unnecessarily complicated. In fact, it is a useful exercise in combining the definition of the fractional conversion of a component with the multiple reactions occurring. Given the fact that we assume Reaction 1 has a 30°C approach ($T = 820$ K) and Reaction 2 is essentially at equilibrium ($T = 850$ K), we can calculate the required K values, and the pressure is fixed at 26.9 atm. The two non-linear equations generated can be solved using the SOLVER function in Microsoft Excel with appropriate constraints (see Appendix 2). It would be useful to check that the quoted answers can be reproduced. As has been said elsewhere, there are a number of solution methods that could be attempted.

The generated values of z_{CH4} and z_{CO} are 0.781 and 0.461, respectively. Thus, the output values from the primary reformer can then be fixed.

Using the conversions calculated, we can use the analysis in Table 1.3 to produce the output values given in Table 1.4.

TABLE 1.4

Results for Output from the Primary Converter Analysis

Component	Initial Moles (kmol)	Moles after Reaction 1 (kmol)	Moles after Reaction 2 (kmol)
CH_4	94.68	21.6	21.6
N_2	1.40	1.4	1.4
CO		77.01	41.51
H_2O	363.3	286.6	251.1
CO_2			35.5
H_2		231.02	266.5
Total			617.62

It will be noted that a significant quantity of hydrogen has been produced, but correspondingly, there is significant unreacted methane, an unwelcome quantity of carbon monoxide and also steam that will have to be considered within the stoichiometry of subsequent processing.

1.7 PRIMARY REFORMER ENERGY BALANCE

The primary reformer is overall endothermic. The pre-processed gas entering the reformer is considered to be at 673 K (400°C), and as has been stated, the exit temperature is fixed at 1123 K (850°C).

An energy balance based on these temperatures is carried out to find the amount of fuel that will supply the energy requirements of the reactor. The analysis carried out here requires enthalpy values. These can be calculated in a number of ways with the outcomes for the fuel requirement being (as they should be!) within reasonable agreement.

In theory, we can carry out an appropriate energy balance over the primary reformer based on the mass balance figures obtained for the reformer.

It is self-evident that to do the energy balance, values of data relevant to energy transfer are required. Appendix 3 outlines the data that were available to carry out the energy balance. These data were enthalpy values for each of the species in the primary reformer from different sources. As has been previously pointed out and is repeated, the purpose of Appendix 3 is not to give a detailed presentation of the enthalpy property, but some basics are presented. The various data sources are reported. It was not considered relevant that differences in the data should be highlighted. The important element was that the data from different sources were in reasonable agreement and could be presented as giving the energies involved in the processes described. As will be seen in the energy balance carried out, the data given by Graham M Hampson (Private Communication) were used and the figures calculated are used as the basis in subsequent calculations. The relevant polynomial constants are given in Table A3.3 in Appendix 3.

To carry out a calculation for the relevant enthalpies, the input and output operational temperatures are required. Normally, if these are not quoted in the text, the values fixed can be found on the block diagram.

Once the appropriate values of enthalpy were available from the Hampson polynomials, the simple energy balance, based on the assumption that the reactor could be considered adiabatic, was carried out (Table 1.5):

$$Input\ Energy = Output\ Energy$$

TABLE 1.5
Primary Reformer Energy Balance

	Input (kmol)	Output (kmol)	Input Specific Enthalpy (kJ/kmol)	Input Enthalpy (kJ)	Output Specific Enthalpy (kJ/kmol)	Output Enthalpy (kJ)
CH_4	94.68	21.6	−3.9715 + 04	−3.7602E + 06	−9.4027E + 03	−2.0310E + 05
C_2H_6	3	0.000	−4.4367 + 04	1.331E + 05	−	−
C_3H_8	0.5	0.000	−4.8422 + 04	−2.4211E + 04	−	−
C_4H_{10}	0.4	0.000	−5.1880 + 04	−2.0752E + 04	−	−
N_2	1.4	1.400	−2.0738 + 04	2.9033E + 04	3.3563E + 04	4.6988E + 04
H_2O	363.3	251.1	−2.1600E + 05	−7.8473E + 07	−1.9799 E + 05	−4.9715E + 07
CO	0	41.51	−	−	−7.9390E + 04	−3.2955E + 06
H_2	0	266.52	−	−	3.4231E04	9.1233e + 06
CO_2	0	35.5	−	−	−3.5034E05	−1.2437E + 07
			Total input	−8.2508E + 07	Total output	−5.6716E + 07
					Output−input	2.5792E + 07

For the purposes of calculating the fuel requirements, the value of $2.5792E + 07\,kJ$ will be used.

1.8   ENERGY SUPPLY FOR PRIMARY REFORMER BY BURNING NATURAL GAS

Energy is provided by burning a suitable fuel in a furnace box through which the catalyst tubes pass. For the purposes of this study, the same natural gas as used for the feedstock will be employed. This will be assumed to enter the burner at 25°C. Combustion air is usually heated and will be supplied at 200°C, and flue gases will necessarily leave at temperatures higher than the reformer gas temperature, usually at about 950°C–1000°C. For the purposes of this calculation, we will use 950°C.

1.9   COMBUSTION MASS BALANCE

In considering the combustion, the combustion stoichiometry needs to be considered. Certain estimates are included in this calculation. We will assume that the air is supplied with an excess of 5%. The furnace will be under vacuum so that air leakage of 10% into the furnace will be assumed; this air will be at a temperature of 10°C.

The following stoichiometric equations will be assumed (the gas used will not have been desulphurised):

$$CH_4 + 2O_2 \rightarrow CO_2 + 2H_2O$$

$$C_2H_6 + \frac{7}{2}O_2 \rightarrow 2CO_2 + 3H_2O$$

$$C_3H_8 + 5O_2 \rightarrow 3CO_2 + 4H_2O$$

$$C_4H_{10} + \frac{13}{2}O_2 \rightarrow 4CO_2 + 5H_2O$$

$$H_2S + \frac{3}{2}O_2 \rightarrow H_2O + SO_2$$

In analysing the combustion of the natural gas as an energy source, there are a number of calculated figures quoted. These figures are not difficult to obtain in themselves, but it is vital that each number is checked and understood so that the mass and energy balances make sense. The learning process depends on the reader doing this; otherwise, the labour involved in the calculations is wasted.

We can draw up the following table (Table 1.6) based on 100 kmol of natural gas and the stoichiometry of the combustion equations based on 100% burning:

TABLE 1.6
Product Amounts from Combustion of Natural Gas assuming 100% Conversion

Species	kmol in Gas	O_2 Required	CO_2 Produced	H_2O Produced	SO_2 Produced
CH_4	94.68	189.36	94.68	$94.68 \times 2 = 189.36$	
C_2H_6	3.0	10.50	$2 \times 3 = 6.00$	$3 \times 3 = 9.00$	
C_3H_8	0.5	2.50	$3 \times 0.5 = 1.50$	$4 \times 0.5 = 2.00$	
C_4H_{10}	0.4	2.60	$4 \times 0.4 = 1.60$	$5 \times 0.4 = 2.00$	
H_2S	0.02	0.03		0.02	0.02
N_2	1.4	0			
Total	100.00	205.0	103.78	202.38	0.02

If we supply the air at 5% excess, then the extra amount of O_2 supplied will be:

$$205 \times 0.05 = 10.25 \text{ kmol.}$$

Thus, the O_2 supplied to the burner should be:

$$205.0 + 10.25 = 215.25 \text{ kmol.}$$

If we also take a leakage of air into the furnace based on the original O_2 calculation and using the 10% leakage already stated, then we get a leakage of O_2 into the furnace of:

$$205.0 \times 0.10 = 20.5 \text{ kmol.}$$

We can take the total O_2 based on 100 kmol of fuel as:

$$205.0 + 10.25 + 20.5 = 235.75 \text{ kmol.}$$

Taking the composition of the air as 21% v/v O_2 and 79% v/v N_2, we can calculate the amount of N_2 as:

$$235.75 \times 79/21 = 886.9 \text{ kmol} + 1.4 \text{ kmol} \quad \text{(from the fuel).}$$

1.10 COMBUSTION ENERGY BALANCE

Thus, for 100 kmol of fuel we can do an enthalpy balance to calculate the energy available from burning the fuel. We note that of the total oxygen input, 215.25 kmol is supplied at 200°C and 20.5 kmol at 10°C.

1.10.1 INPUT ENTHALPY

TABLE 1.7
Input Enthalpies of Entering Natural Gas and Air Streams

		Enthalpy of Fuel	
Species	kmol in Gas	kJ/kmol at 25°C	kJ
CH_4	94.68	−5.6719E+04	−5.3702E+06
C_2H_6	3.0	−5.6296E+04	−1.6889E+05
C_3H_8	0.5	−6.7061E+04	−3.3531E+04
C_4H_{10}	0.4	−9.0014E+04	−3.6006E+04
H_2S	0.02	−1.9960E+04	−3.9900E+02
N_2	1.4	8.6460E+03	1.2104E+04
Total	100.00		−5.5969E+06

		Enthalpy of Entering Air/100 kmol of Fuel	
Species	kmol in Gas	kJ/kmol at 200°C	kJ
O_2	215.25	1.3913E+04	2.9948E+06
N_2	215.25×79/21 = 809.75	1.3796E+04	1.1171E+07
		kJ/kmol at 10°C	
O_2	20.5	8.1340E+03	1.6675E+05
N_2	20.5×79/21 = 77.12	8.2270E+03	6.3447E+05
Total			1.4967E+07

Check out the enthalpy values using the polynomial constant values used in Appendix 3. This will be required in the subsequent energy balances to be carried out. Although these calculations and results are not carried out here, it might be a useful exercise to check out enthalpy values (Table 1.7) from other sources as given in Appendix 3.

Thus, the total enthalpy of the input streams is:

$$-5.5969E+06+1.4967E+07=9.3701E+06 \text{ kJ}$$

1.10.2   OUTPUT ENTHALPY

The amounts of CO_2 and H_2O produced have all been previously calculated assuming complete combustion.

The amount of N_2 associated with the input oxygen can also be calculated as:

$$809.75+77.12=886.87 \text{ kmol}$$

(See Table 1.7; we note that 1.4 kmol has come through from the fuel)

If we assume that the stoichiometric amount of O_2 is burned with the excess O_2 and leakage O_2 passing into the flue gas, then we can get the amount of O_2 in the flue gas as:

$$10.25+20.5=30.75 \text{ kmol}$$

$$\left(\text{i.e. Excess } O_2 \text{ already calculated} + \text{Leakage } O_2 \text{ already calculated}\right)$$

Taking the flue gas components, we can attempt to calculate an enthalpy figure for the output gas (Table 1.8):

TABLE 1.8

Calculated Enthalpies for Output Gases from the Combustion

Species	kmol in Gas	kJ/kmol at 950°C	kJ
CO_2	103.78	$-3.3754E+05$	$-3.5030E+07$
H_2O	202.38	$1.9649E+05$	$-3.9766E+07$
N_2	$886.87+1.4=888.27$	$3.6459E+04$	$3.2385E+07$
O_2	30.75	$3.9190E+04$	$1.2051E+06$
Total	1225.18		$-4.1205E+07$

If we then carry out an energy balance for the fuel stream,

$$\text{Energy supplied} / 100 \text{ kmol of Fuel} = \text{Energy Output} - \text{Energy Input}$$

$$\text{Energy supplied} = -4.1205E+07-9.3701E+06 = -5.0575E+07 \text{kJ}$$

We have already calculated that the energy requirement of the reformer is 2.5792E + 07 kJ (see Table 1.5).

For 100 kmol of fuel, the energy supplied is −5.0575E + 07 kJ.

The energy required is 2.5792E + 07; thus, the amount of fuel required is:

$$100 \times 2.5792E + 07/5.0575E + 07 = 51.0 \, kmol$$

Assume that, with heat losses, the fuel requirement will be:

$$51.0 \times 1.02 = 52.0 \, kmol$$

$$Combustion \, air = 215.25 \left(amount \, of \, O_2 \, already \, calculated\right) + 809.75$$

$$\left(amount \, of \, N_2 \, already \, calculated\right) = 1025 \times 52.0/100 = 533 \, kmol$$

$$Leakage \, air = 20.5 + 77.12 = 97.62 \times 52/100 = 50.8 \, kmol$$

$$Flue \, gas = 1225.18 \times 52/100 = 637.1 \, kmol$$

The input and output temperature and pressure stream conditions were taken as follows:

Input
Temperature = 400°C
Pressure = 32.06 bar

Output
Temperature = 850°C
Pressure = 27.25 bar.

A reasonable pressure drop for such a reformer is assumed. This can be confirmed with a suitable design-based calculation.

SUMMARY OF IMPORTANT PARAMETERS CALCULATED FOR THE PRIMARY REFORMER

Mass output from the primary reformer which required the use of calculated equilibrium constants.

At the reformer output temperature of 850°C, K values were calculated using the following:

METHANE–STEAM REACTION

$$\ln K_A = 30.42 - \left(\frac{27,106}{T}\right)$$

WATER–GAS SHIFT REACTION

$$\ln K_{P_2} = -3.798 + \left(\frac{4160}{T}\right)$$

The fractional conversions for the two reactions were calculated as 0.781 and 0.461, respectively. The output components kmol are summarised in Table 1.4. The required energy input to the reformer has been calculated as:

$$2.5792E + 07\,kJ.$$

The same natural gas as for the feedstock is burned to supply this energy. The energy balance indicates that 52.0 kmol of gas must be burned to supply the required energy.

1.11 SECONDARY REFORMER (F102 ON THE BLOCK DIAGRAM)

REQUIRED

Primary reformer input to secondary reformer.
Information on water–gas shift reaction K values:

WATER–GAS SHIFT REACTION

$$\ln K_{P_2} = -3.798 + \left(\frac{4160}{T}\right)$$

Fixing of amount of air required to supply the nitrogen in the syngas in the equilibrium ratio of

$$H_2/N_2 = 3:1.$$

Fixing the output component masses by solving component mass balances and equilibrium equations simultaneously.
Confirming the energy input and output streams balance.
Understanding some of the implications of the secondary reformer outputs for further operations in the block diagram.

1.12 SECONDARY REFORMER MASS BALANCE

Negligible methane should enter the ammonia synthesis loop; hence, the methane content of the entering stream to this loop should be reduced. This is usually achieved in a secondary reformer that would normally operate 150–250°C in excess of the exit temperature from the primary reformer. With such a temperature increase, the methane content is expected to reduce to <0.5%. This higher temperature can be achieved by using the energy released in the exothermic reaction that involves the combustion of some of the methane in the stream from the primary reformer. The oxygen necessary for the combustion is supplied by injection of combustion air. The amount of air injected has to be fixed by the requirement that the amount of nitrogen entering the ammonia synthesis loop is in the ratio $N_2/H_2 = 1:3$. The secondary reformer is obviously supplying the bulk of the necessary nitrogen for the syngas.

In the environment of the secondary reformer, it is obvious that as well as the combustion reaction with oxygen, there could be further reforming of the methane by the methane–steam reaction and the possibility of the water–gas shift reaction occurring. It is important in carrying out a mass balance that these possibilities are recognised and the appropriate calculations carried out.

The input to the secondary reformer has been fixed by the calculations carried out on the primary reformer. To fix the oxygen and hence combustion air requirements, certain assumptions can be made.

We assume that certain important components can be represented by the following:

Component	Symbol
Carbon monoxide, CO	M
Carbon dioxide, CO_2	D
Nitrogen, N_2	N
Hydrogen, H_2	H
Water (steam), H_2O	W

We can assume that the water–gas shift reaction takes place in the secondary reformer:

$$CO + H_2O \rightarrow CO_2 + H_2$$

At the end of the synthesis gas production, we try to process to ensure that the CO content is negligible; the implication of this is that at the exit from the secondary reformer, the ratio $(CO + H_2)/N_2$ has to be 3/1. This enables us to set up balances that can be used to calculate the amounts of each component in the secondary reformer exit.

If we take the exit component amounts from the primary reformer, we can then write the following atom/molecule balances and other relevant equations (Table 1.9):

In terms of our unknowns, we have indicated five; we need to identify five independent equations to obtain quantities for each of these components at the reformer exit. Thus:

$$\text{C-atom Balance}: 97.81 = M + D \tag{1.1}$$

$$\text{O-atom Balance}: 363.21 + 0.532N = M + 2D + W \tag{1.2}$$

$$\text{H-atom Balance}: 1119.24 = 2H + 2W \tag{1.3}$$

TABLE 1.9
Atom Balance for Components from the Primary Reformer

Component	kmol	C	O	H	Symbol and Quantity at Exit
CH_4	21.0	21.0	0	84.0	
CO	41.51	41.51	41.51	0	M
CO_2	35.30	35.30	70.60	0	D
H_2	266.52	0	0	533.04	H
H_2O	251.1	0	251.1	502.2	W
N_2	1.4	0	0	0	N
Air (O_2)	$\dfrac{21}{79}N$	0	$\left(\dfrac{21}{79}X2\right)N$	0	
Total		97.81	$363.21+0.532N$	1119.24	

As stated above, the ratio $(CO+H_2)/N_2$ has to be 3/1, and we can write:

$$M + H = 3N \tag{1.4}$$

The fifth equation in these calculations is provided by considering the K_p value for the water–gas shift reaction. As stated previously, the secondary reformer normally operates at a temperature 150°C–250°C in excess of the exit temperature from the primary reformer. In this case, we will assume a temperature of 1080°C for the exit temperature. We can write K_p as:

$$K_P = \frac{H \times D}{M \times W}$$

Because we have fixed an exit temperature, we can compute a value of K_p from the equation previously quoted for this reaction:

1.12.1 WATER–GAS SHIFT REACTION

$$\ln K_P = -3.798 + \left(\frac{4160}{T}\right)$$

where T is in K.

At a temperature of 1080°C = 1353 K, we can compute a value of K_p as:

$$K_P = \frac{H \times D}{M \times W} = 0.485 \tag{1.5}$$

These five equations can be solved simultaneously (Appendix 4) and a value for M obtained.

A brief description of a method using SOLVER in an Excel spreadsheet is described in Appendix 4. A classic substitution/elimination approach is also illustrated.

We find the value of M to be 64.41 kmol. Thus, we know the amount of CO exiting the secondary reformer.

The amounts of the other components exiting the reformer can then be calculated using the appropriate equations for H, M, W and D, and this is also illustrated in Appendix 4.

We can make an estimate of the CH_4 that has slipped through the process. If we assume that the methane–steam reaction can also take place in the reformer at 1080°C, then we can calculate the K_P value for the reaction at this temperature from the equation used previously:

1.12.2 Methane–Steam Reaction

$$\ln K_P = 30.42 - \left(\frac{27,106}{T}\right)$$

With a 30°C approach for this reaction, we take the temperature as 1050°C = 1050 + 273 = 1323 K.

We can calculate a value of K_P as 20572.6.

We are operating at a pressure of 25.77 bar.

We write for K_P:

$$K_P = \frac{n_{CO}(n_{H_2})^3}{n_{CH_4}n_{H_2O}}\left(\frac{P}{\sum n}\right)^2$$

where n is the number of moles for each component and $\sum n$ is the total moles in the equilibrium mixture. If we consider the total kmol of all the components except methane in the reformer exit stream, we get a value of 769.82 kmols. We can then express the number of kmols of CH_4 in the outlet as n_{CH4} and the total moles in the outlet as $769.82 + n_{CH4}$. If we then write an equation:

$$20,572.6 = \frac{64.41 \times 269.17^3}{n_{CH_4} \times 290.45}\left(\frac{25.77}{n_{CH_4} + 769.82}\right)^2$$

If we solve for n_{CH4}, we obtain a value of $n_{CH4} = 0.24$ kmol.

We can write a summary table (Table 1.10).

1.13 SECONDARY REFORMER ENERGY BALANCE

The main principles and treatment of enthalpy data were set out in the treatment of the energy balance for the primary reformer.

From Table 1.5, we can obtain the enthalpy of the input stream to the secondary reformer (exit from the primary reformer) as $-5.6716E + 07\,kJ$. It is assumed that preheated air will be delivered at a temperature of 300°C.

The amount of air entering is 111.19 kmol N_2 and 29.5 kmol O_2.

TABLE 1.10

Summary of Mass Balance on the Secondary Reformer

Component	kmol at Input	kmol at Exit
CH_4	21.0	0.24
CO	41.51	64.41
CO_2	35.30	33.2
H_2	266.52	269.17
H_2O	251.1	290.45
N_2	1.4	$111.19+1.4=112.59$
Air (O_2)	$\frac{21}{79}N = 29.5$	
Total		770.06

Thus, the enthalpy of the air entering can be calculated by finding enthalpies for N_2 and O_2 using the polynomial data from Table A3.5. At 573 K, the enthalpy of N_2 is $1.705E + 04\,kJ$ and O_2 is $1.7069E + 04\,kJ$.

The enthalpy for the air stream is:

$$111.19 \times 1.705E + 04 + 29.5 \times 1.7069E + 04 = 2.3993E + 06 \text{ kJ}$$

Thus, the total input enthalpy is:

$$-5.6716E + 07 + 2.3993E + 06 = -5.4317E + 07\,kJ$$

The output data can be calculated using the data from Table A3.5 for an output temperature of 1080°C.

TABLE 1.11

Enthalpy Data for Output Stream from the Secondary Reformer

Component	kmol at Exit	Enthalpy (kJ/kmol)	Enthalpy (kJ)
CH_4	0.24	$8.9690E+03$	$2.1526E+03$
CO	64.41	$-6.8424E+04$	$-4.4072E+06$
CO_2	33.2	$-3.3024E+05$	$-1.0964E+07$
H_2	269.17	$4.2055E+04$	$1.1320E+07$
H_2O	290.45	$-1.9165E+05$	$-5.5665E+07$
N_2	$111.19+1.4=112.59$	$4.0152E+04$	$4.5207E+06$
Total	770.06		$-5.5193E+07$

The assumption is that we have a steady state energy balance, i.e.:

Energy Input to the Secondary Reformer = Energy Out from the Secondary Reformer

In this case, we are essentially assuming an adiabatic operation with no heat leaks assumed.

If we examine our figures:

$$\text{Input Stream Enthalpies} = \text{Output Stream Enthalpies}$$

$$-5.4317\text{E}+07\,\text{kJ} = -5.5193\text{E}+07\,\text{kJ}$$

Within round-off errors, this is considered to be acceptable (difference $< 2\%$).

In terms of the exit gas from the secondary reformer, we note the high stream temperature (1080°C). This stream can be used to raise steam within the process, and this is achieved within the *reformed gas boiler (Item E102 on the block diagram)*. For efficient energy usage, a calculated amount of steam is raised in E102, and then, further energy exchange occurs in the *heat interchanger (Item E106 on the block diagram)*. In this item, the high-temperature stream exchanges energy with a low-temperature gas stream exiting the *absorber (Item V105 on the block diagram)*. This item removes the remaining CO_2 from the ammonia synthesis gas stream. The temperature of this stream will fix the temperature of the gas stream exiting the *heat interchanger (E106)* prior to entry into the *methanator (Item V107 on the block diagram)*.

In terms of the current mass and energy balance calculations, the temperature of the gas stream entering the *high-temperature shift converter 1 (Item V102 on the block diagram)* has to be considered. Table 1.11 shows that there is still a considerable amount of carbon monoxide present in the synthesis stream ($64.41/770.06 = 0.084$, i.e. 8.4 mole %); this has to be removed prior to entering the ammonia synthesis loop to preserve the catalyst activity. This removal is carried out using the water–gas shift reaction that has been discussed previously when considering the primary reformer operation.

At 298 K, we can write:

$$CO + H_2O \leftrightarrow CO_2 + H_2 \quad \Delta H = -41\,\text{kJ}$$

Essentially, the action we require, the removal of CO, is being carried out with the production of further hydrogen. (We tried to adjust for this in our N_2/H_2 ratio calculations in the secondary reformer.) The removal of the CO_2 has to then be considered.

It can be seen from the reaction enthalpy figure quoted with the stoichiometric equation that the reaction is exothermic. Thus, we face the classic problem. Thermodynamically, we should reduce the temperature to increase the thermodynamic conversion of the CO. Kinetically, we should increase the temperature so that the final thermodynamic equilibrium is approached quickly. The necessary compromise is achieved by using appropriate catalysts and arranging the temperature conditions for the reaction. In this example, a *high-temperature shift converter and a low-temperature shift converter* are used. As described previously, the shift converters use the water–gas shift reaction to remove carbon monoxide before the synthesis gas enters the ammonia production reactor. The water–gas shift reaction is exothermic and thermodynamically should be carried out at a low temperature. In the current set-up, this is carried out in a two-stage high-temperature shift reactor followed by a low-temperature shift converter. The high-temperature stages allow favourable kinetic conditions to operate with suitable catalysts. This stage can be improved by the use of two beds of catalyst. The product gases can then be allowed to enter a lower temperature catalyst bed where the more favourable thermodynamic conditions allow the water–gas shift to produce further carbon monoxide conversion as it approaches equilibrium. The kinetic aspects are still dealt with by the shift reaction taking place in the presence of the chosen catalyst.

Some plants using this synthesis route now only employ a shift converter with no high- and low-temperature differentiation. Improved catalysts and catalyst technology usually mean that satisfactory product amounts are obtained. For the purposes of illustrating the relevant mass and energy balances, this study calculates using two catalyst beds in a high-temperature shift converter followed by a single-bed low-temperature shift converter.

The gas from the *interchanger (E106)* enters the *high-temperature shift converter* which contains an iron–chromium catalyst that is reported to be active in the range 300°C–500°C. From the balances carried out/to be carried out, the calculations for this converter will be done with an entering temperature of 389°C. Because the reaction is exothermic, the temperature rises. Typically, for the conditions considered here, a temperature rise of 65°C–80°C can be expected. This work assumes a rise of 70°C. This gives an exit temperature of 459°C. To further reduce the CO content using this catalyst, the exit stream at 459°C is cooled and the stream energy utilised in *converted gas boiler A (Item E103A on the block diagram)*. The exit gas stream is then put in to a second bed containing the same catalyst, and the CO content is further reduced. This is the *high-temperature shift converter 2 (Item V102 on the block diagram)*.

With the discovery of a Cu/Zn/alumina catalyst, the remaining CO can be significantly reduced at a lower temperature. This lower temperature is achieved by utilising the stream enthalpy for the high-temperature shift converter exit stream in the *converted gas boiler B (Item E103B on the block diagram)* and further lowering the temperature using a *quench cooler (Item M101 on the block diagram)*. The stream from the quench cooler then enters the *low-temperature converter (Item V103 on the block diagram)*.

SUMMARY OF IMPORTANT PARAMETERS CALCULATED FOR THE SECONDARY REFORMER

The component masses have been calculated by solving five simultaneous equations. The results are summarised in Table 1.10. This fixes the amount of hydrogen and nitrogen needed for subsequent processing.

The input and output energies should balance for the mass changes made. The values of the input and output enthalpies were calculated as follows:

$$-5.4317E+07\,kJ = -5.5193E+07\,kJ$$

This is considered acceptable.

1.14 REFORMED GAS BOILER (ITEM E102 ON THE BLOCK DIAGRAM)

REQUIRED

Calculation of the energy available from the secondary reformer output stream to raise stream in a boiler.

The energy available requires a calculation of the input and output enthalpies based on input and output temperatures.

The difference between the input and output energies indicates the amount of energy available for stream raising.

Enthalpies of vaporisation (latent heat) taken from appropriate thermodynamic tables (steam tables) allow calculation of the amount of steam that can be raised.

The input stream to this boiler is the exit stream from the secondary reformer. The component quantities, stream enthalpies, temperature and pressure for this stream are known and shown in Table 1.12.

Temperature: 1080°C

Pressure: 26.1 bar

Total kmol: 770.06

Stream enthalpy: $-5.5193E + 07\,kJ$

The energy available in this stream is used in the boiler to raise high-pressure steam. The pressure of this steam is plant determined and in this case is generated at a pressure of 62 bar. The exit temperature of the gas stream is determined by the temperature required in the *interchanger (Item E106)*. There is a stream from the interchanger to the *methanator (Item V107)*; the operation of this is discussed later in these calculations. The suggested operating temperature range for this operation is 250°C–350°C. The temperature finally used for calculation in this operation is 326°C at a pressure of 21.8 bar. The calculations carried out fix a temperature for the input gas stream to the interchanger (the output stream from the reformed gas boiler (E102)); this temperature is 535°C and this is used in the calculations related to the *reformed gas boiler (E102)*. We again look at Table 1.12.

TABLE 1.12
Input Mass to Reformed Gas Boiler and Component and Stream Enthalpies

Component	kmol at Exit	Enthalpy (kJ/kmol)	Enthalpy (kJ)
CH_4	0.24	8.9690E + 03	2.1526E + 03
CO	64.41	−6.8424E + 04	−4.4072E + 06
CO_2	33.2	−3.3024E + 05	−1.0964E + 07
H_2	269.17	4.2055E + 04	1.1320E + 07
H_2O	290.45	−1.9165E + 05	−5.5665E + 07
N_2	111.19 + 1.4 = 112.59	4.0152E + 04	4.5207E + 06
Total	770.06		−5.5193E + 07

We have the data on the input stream, and we now need to fix the enthalpy of the output stream; obviously, there is no change in the mass balance.

We can use the constants given in Table A3.3 (Appendix 3) in the polynomial equation to find component enthalpies at 535°C.

TABLE 1.13

Output Mass from Reformed Gas Boiler and Component and Stream Enthalpies

Component	kmol at Exit	Enthalpy (kJ/kmol)	Enthalpy (kJ)
CH_4	0.24	−3.1586E+04	−7.5810E+03
CO	64.41	−8.9134E+04	−5.7411E+06
CO_2	33.2	−3.6098E+05	−1.1984E+07
H_2	269.17	2.4007E+04	6.4620E+06
H_2O	290.45	−2.1170E+05	−6.1487E+07
N_2	111.19+1.4=112.59	2.4193E+04	2.7239E+06
Total	770.06		−7.0035E+07

Thus, the energy transfer over the boiler is:

$$= \text{Energy Out} - \text{Energy In}$$

$$= -7.0035E+07 - (-5.5193E+07) = -1.4842E+07 \, \text{kJ}$$

This is an energy output available to the BFW. If we factor in a small heat loss to the atmosphere – say 1% – then the energy available is:

$$-1.4842E+07 \times 0.99 = -1.4694E+07 \, \text{kJ}.$$

We have stated the steam down pressure as 62 bar. Essentially, we are using the gas stream energy to supply latent heat (enthalpy of vaporisation) to the water.

From steam tables:

Latent heat of water at 60 bar = 1570 kJ/kg
Latent heat of water at 65 bar = 1538 kJ/kg

Using linear interpolation:

$$\left(\frac{62-60}{65-60}\right) \times (1538-1570) + 1570 = 1557.2 \, \text{kJ/kg}$$

Thus, the steam raised in the boiler is:

$$\frac{1.4594E+07}{1557.2} = 9436 \, \text{kg}$$

SUMMARY OF IMPORTANT PARAMETERS
CALCULATED FOR REFORMED GAS BOILER

Enthalpies of input and output streams to the boiler are given in Tables 1.12 and 1.13.

Energy available to reformed gas boiler is $-1.4694E+07$.

Using appropriate latent heat data, the amount of stream raised in the boiler is 9436 kg.

REFERENCE

1. Richard M Felder, Ronald W Rousseau, December 2004, *Elementary Principles of Chemical Processes*, Wiley Inc, USA.

2 Shift Converters and Stream Energy Interchange

2.1 SHIFT CONVERTERS

The gases exiting the reformed gas boiler contain a significant amount of carbon monoxide (13.43% on a water-free basis). Most of this CO will, potentially, be converted into carbon dioxide (CO_2). Inspection of the block diagram indicates that later in the process for syngas production, a *methanator* is inserted. Essentially, this device is necessary to remove the carbon oxides (CO and CO_2) from the synthesis gas. Before this final removal stage is reached, the shift conversion reaction is carried out to lower the high CO level in the stream from the reformed gas boiler. The water–gas shift reaction was encountered in calculations on the primary and secondary reformers. The use of shift converters is important in reducing the CO stream content and producing hydrogen. A continued CO presence will consume hydrogen in the methanator, and because this process uses a reversed methane–steam reaction, methane can be produced in the *methanator* and feed unwanted inerts into the final synthesis gas.

As the production of this synthesis gas has evolved, the shift reaction stages have changed. With more efficient catalysts available, the operating conditions and number of shift reactors have changed. In this text, for mass and energy balance purposes, an original layout is presented in which a two-stage shift converter operates at a high temperature followed by a shift converter operating at a lower temperature. The water–gas shift reaction is exothermic, and thermodynamics dictates that it should be carried out at as high a temperature as possible. The higher temperature is expected to favour the reaction rate. The classic compromise of a reaction temperature set at some intermediate level is adopted in a high-temperature shift converter followed by a shift converter operating at a lower temperature, which favours a higher equilibrium conversion and encourages the use of efficient catalysts to produce an acceptable reaction rate.

2.2 HIGH-TEMPERATURE SHIFT CONVERTER (ITEM V102 ON THE BLOCK DIAGRAM)

REQUIRED

Input component masses and enthalpies to high-temperature shift converter bed 1 (Table 2.1).

Calculation of equilibrium constant for water–gas shift reaction in reactor:

$$\ln K_P = \frac{4510.23}{T} - 4.196$$

Calculation of fractional conversion (z) for the first bed.
  Based on z, calculation of the output component masses from the first bed.
  Output stream from Bed 1 becomes input to Bed 2.
  Equilibrium constant calculated for Bed 2.
  Calculation of fractional conversion for Bed 2.
  Energy balance calculated for high-temperature shift beds 1 and 2.

As has been indicated, this converter operates in two stages. The inlet conditions to the first stage are dictated by the temperature of the synthesis gas exiting the *energy interchanger (Item E106 on the block diagram)*.

The *interchanger* is dealt with subsequently and the exit reformed gas temperature from the *interchanger* is assumed to be 389°C at a pressure of 25.3 bar, and this is the feed stream into the *high-temperature shift converter first stage*. The water–gas shift conversion reaction takes place, and the outlet temperature is fixed at 459°C and pressure 24.9 bar. Using the polynomial for the component enthalpies (Table A3.3, Appendix 3), the following table can be drawn up for the first bed of the high-temperature shift converter.

2.3   MASS AND ENERGY BALANCE CALCULATIONS FOR BED 1 OF THE HIGH-TEMPERATURE SHIFT CONVERTER

TABLE 2.1

Input Mass, Component and Stream Enthalpies for Bed 1

Component	Amount (kmol)	Specific Enthalpy (kJ/kmol)	Enthalpy
H_2	269.17	1.9463E+04	5.2389E+06
CO	64.41	−9.4374E+04	−6.0786E+06
CO_2	33.2	−3.6855E+05	−1.2236E+07
CH_4	0.24	−4.0336E+04	−9.6810E+03
N_2	112.59	1.9765E+04	2.2253E+06
H_2O	290.45	−2.1687E+05	−6.2988E+07
Total	770.06		−7.3848E+07

Note: Bed 1 inlet temperature is 389°C.

The outlet temperature from Bed 1 has been fixed at 459°C. The catalyst used in this bed (iron oxide/chromium oxide) is reported to have an approach to equilibrium (see Appendix 5) of 31°C. To fix the output kmol, we have to consider the water–gas shift reaction:

$$CO + H_2O \leftrightarrow CO_2 + H_2$$

With an approach to equilibrium of 31°C, we can take a K (equilibrium constant) value for this reaction at $459 + 31 = 490$°C, which is 763 K. Using the equation quoted in Appendix 1 for this reaction:

$$\ln K_P = \frac{4510.23}{T} - 4.196$$

At 763 K, we find that K_P is 5.56.

It is relatively easy to then demonstrate, using the figures quoted in Table 2.1, that for a fractional conversion of z for CO, we can write:

$$K_p = 5.56 = \frac{(33.2 + 64.41z)(269.17 + 64.41z)}{(64.41 - 64.41z)(290.45 - 64.41z)}$$

where z is the fractional conversion of CO at the equilibrium temperature of 490°C.

This gives a quadratic equation in z. z can be calculated as 0.715. Taking this figure, we can then calculate the output species moles exiting the first bed of the high-temperature converter and produce the following table:

TABLE 2.2

Output Mass, Component and Stream Enthalpies for Bed 1

Component	Amount (kmol)	Specific Enthalpy (kJ/kmol)	Enthalpy
H_2	315.22	2.1627E+04	6.8173E+06
CO	18.36	−9.1710E+04	−1.6838E+06
CO_2	79.25	−3.6499E+05	−2.8926E+07
CH_4	0.24	−3.6278E+04	−8.7070E+03
N_2	112.59	2.1893E+04	2.4649E+06
H_2O	244.44	−2.1440E+05	−5.2408E+07
Total	770.1		−7.3744E+07

Note: Bed 1 outlet temperature is 459°C.

On the basis that for the first bed, energy input = energy output, the input and output energies agree within less than 0.5%. If there is a problem in simply using the polynomials for an operation with chemical reaction, an examination of the van't Hoff box laid out in Appendix 3 should be carried out. An examination of the component outputs from the first stage indicates that there is still a significant presence of carbon monoxide.

The output temperature (459°C) is considerably higher than input (389°C) as would be expected from the exothermic reaction that has taken place. The entry temperature for the second bed of the shift converter needs to recognise that the reactor is still thermodynamically controlled and temperature needs to be as low (for an exothermic reaction) as can be realistically set. In this example, the inlet temperature to the second bed is set very close to the inlet temperature of the first bed. To achieve this, as can be seen from the block diagram, energy is removed from the first-bed exit stream by suitable heat exchange to raise steam in the *converted gas boiler A, Item E103A* on the block diagram. The energy that can be removed can be quantified by carrying out an energy balance over this item.

2.4　MASS AND ENERGY BALANCE CALCULATIONS FOR BED 2 OF THE HIGH-TEMPERATURE SHIFT CONVERTER

For the purposes of calculation, the second bed will be considered and then calculations carried out on the *converted gas boilers A and B*. The stream exiting *converted gas boiler B* requires further cooling, and from the cooling aspect, there is an argument that the low temperature of the stream entering the *low-temperature shift converter* can be sensibly achieved by using quench cooling and this is in fact used.

TABLE 2.3

Input Mass, Component and Stream Enthalpies for Bed 2

Component	Amount (kmol)	Specific Enthalpy (kJ/kmol)	Enthalpy
H_2	315.22	1.9401E+04	6.1156E+06
CO	18.36	−9.4085E+04	−1.7274E+06
CO_2	79.25	−3.6865E+05	−2.9215E+07
CH_4	0.24	−4.0448E+04	−9.7080E+03
N_2	112.59	1.9704E+04	2.2185E+06
H_2O	244.44	−2.1694E+05	−5.3028E+07
Total	770.1		−7.5646E+07

Note: For the second bed, inlet temperature is 387°C.

The outlet temperature from Bed 2 has been fixed at 397°C. The same catalyst used in Bed 1 (iron oxide/chromium oxide) is employed here and has the same approach to equilibrium of 31°C. To fix the output kmol, we again consider the water–gas shift reaction:

$$CO + H_2O \leftrightarrow CO_2 + H_2$$

With an approach to equilibrium of 31°C, we can take a K (equilibrium constant) value for this reaction at 397 + 31 = 428°C, which is 701 K. As before, using the equation quoted in Appendix 1 for this reaction:

At 701 K, we find that K_p is 9.37.

It is relatively easy to then demonstrate, using the figures quoted in Table 2.2, that for a fractional conversion of z for CO, we can write:

$$K_p = 9.37 = \frac{(79.25 + 18.36z)(315.22 + 18.36z)}{(18.36 - 18.36z)(244.44 - 18.36z)}$$

where z is the fractional conversion of CO at the equilibrium temperature of 428°C.

This gives a quadratic equation in z.

z can be calculated as 0.331. Taking this figure, we can then calculate the output species moles exiting the second bed of the high-temperature reformer and produce the following table (Table 2.4):

TABLE 2.4

Output Mass, Component and Stream Enthalpies for Bed 2

Component	Amount (kmol)	Specific Enthalpy (kJ/kmol)	Enthalpy
H_2	321.30	1.9710E+04	6.3328E+06
CO	12.28	−9.3759E+04	−1.1514E+06
CO_2	85.33	−3.6815E+05	−3.1414E+07
CH_4	0.24	−3.9885E+04	−9.5720E+03
N_2	112.59	2.0008E+04	2.2527E+06
H_2O	238.36	−2.1659E+05	−5.1625E+07
Total	770.1		−7.5615E+07

Note: Bed 2 outlet temperature is 397°C.

The inlet and outlet energies agree within less than 0.5%.

The analysis for the component masses out of the second bed of the *high-temperature shift converter* would now allow this stream to become the input stream to *the low-temperature shift converter* to allow the carbon monoxide level to be taken to acceptable process levels. Before the low-temperature shift converter is considered, it is necessary to carry out energy management calculations related to boilers E103A and E103B on the block diagram.

SUMMARY OF IMPORTANT PARAMETERS CALCULATED FOR HIGH-TEMPERATURE WATER–GAS SHIFT CONVERTER

Equilibrium constants for the water–gas shift reactor beds 1 and 2:

For Bed 1, $K_p = 5.56$
For Bed 2, $K_p = 9.37$

Both these values assume an approach to equilibrium (Appendix 5) of 31°C for the catalyst used.

Fractional conversion value (z) for Bed 1 $z = 0.715$
Fractional conversion value (z) for Bed 2 $z = 0.331$
Calculation of input and output enthalpies for both beds (Tables 2.1–2.4)
Calculations show that the energies balance across the beds.

2.5   CONVERTED GAS BOILERS (ITEMS E103A AND E103B) AND THE QUENCH COOLER (ITEM M101 ON THE BLOCK DIAGRAM)

REQUIRED

Previously calculated mass and energy data relating to the input and output streams for the two beds of the high-temperature shift converter (HTS).

Doing the energy balances over the two converted gas boilers to calculate the energy available for steam raising.

Using the energy available calculation of the amount of steam that can be raised in the boilers. Using the output from converted gas boiler B, the output temperature from the quench cooler is used to fix the temperature of the stream exiting the quench cooler.

We can summarise the input and output conditions for the two boilers as follows (Table 2.5):

TABLE 2.5

Input and Output Conditions for Converted Gas Boilers

	E103A (Block Diagram)		E103B (Block Diagram)	
	Inlet	Outlet	Inlet	Outlet
Temperature (°C)	459	387	397	288
Pressure (bar)	24.9	24.6	24.2	23.9

In terms of plant services, using the energy available from the high-temperature shift converter streams, steam will be generated at 62 bar pressure with a corresponding saturation temperature of 278°C.

2.6   ENERGY BALANCES FOR CONVERTED GAS BOILERS A AND B AND THE QUENCH COOLER

If we consider the energy available for *Item E103A*, then we consider the difference in energy between the input stream to the boiler *(output stream from the first bed of the high-temperature shift converter, Item V102; stream enthalpy −7.3744E + 07 kJ)* and the output stream from *boiler E103A (which is the input stream to the second bed of the high-temperature shift converter, Item V102; stream enthalpy already calculated as −7.5615E + 07 kJ).*

Thus, the energy available for steam raising in E103A is

$$(-7.5615E + 07) - (-7.3744E + 07) = -1.8710E + 06 kJ$$

If we assume a realistic heat loss of 1%, then we work on a basis of $-1.8523E+06$ kJ *available for steam raising.*

If we consider the energy available from *converted gas boiler B (Item E103B)*, then we consider the difference between the input stream to the boiler (which is the output stream from the second bed of the high-temperature shift converter, Item V102; enthalpy already calculated as $-7.5615E+07$ kJ) and the output stream from the boiler (Item E103B), the energy content of which is now calculated.

The table below (Table 2.6) is based on the calculated mass balance for the outlet from converted gas boiler B (Item E103B) temperature 288°C

TABLE 2.6
Outlet Masses and Enthalpies for Converted Gas Boiler B

Component	Amount (kmol)	Specific Enthalpy (kJ/kmol)	Enthalpy
H_2	321.30	1.6391E+04	5.2664E+06
CO	12.28	-9.7247E+04	-1.1942E+06
CO_2	85.33	-3.7080E+05	-3.1640E+07
CH_4	0.24	-4.5710E+04	-1.0970E+04
N_2	112.59	1.6684E+04	1.8785E+06
H_2O	238.36	-2.2037E+05	-5.2528E+07
			-7.8229E+07

Thus, the energy available for steam raising in *Item E103B* is:

$$(-7.8229E+07)-(-7.5615E+07)=-2.6140E+06 \text{ kJ}$$

Assuming again that we have a realistic heat loss of 1%, then we work on a basis of $-2.5879E+06$ kJ.

Taking the enthalpy of vaporisation of steam at 62 bar (latent heat) to be **1557 kJ/kg** as supplied in steam tables [1], we can then calculate the following:

For the *converted gas boiler A (Item E103A),* the amount of steam raised is:

$$1.8523E+06/1557 = 1190 \text{ kg}$$

For the *converted gas boiler B (Item E103B),* the amount of steam raised is:

$$2.5879E+06/1557 = 1662 \text{ kg.}$$

This gives a total of steam raised as $1190+1662=2852$ kg, 2.852 tonnes.

The low-temperature shift converter is normally expected to operate in the temperature range 200°C–250°C. After raising steam in the converted gas boilers, the lower grade energy in the stream entering the shift converter is normally processed by lowering the temperature using quench cooling in which a controlled stream of water is allowed to directly enter the process gas stream.

For the purposes of this calculation, the water entering in the **quench cooling** is considered to be 0.094 kmol/kmol of dry process gas.

Using the component amounts shown in Table 2.6, we can calculate the kmol of dry gas (i.e. excluding the kmol of water from the stream) quantity as 531.74 kmol.

On that basis, we will add:

$$531.74 \times 0.094 = 50.0 \, \text{kmol water, which is } 50.0 \times 18 = 900 \, \text{kg}.$$

Using Table 2.6, we have the following:

Water stream quantity at inlet to *quench cooler* is 238.36 kmol.

Now, the water stream at outlet from *quench cooler* is:

$$238.36 + 50.0 = 288.36 \text{ kmol}$$

We now need to carry out an energy balance over the *quench cooler* to fix the outlet temperature from this cooler which is the inlet temperature to the *low-temperature shift converter*. We can fix a reasonable specific enthalpy for the quench stream. In practice, the water for the quench can be taken from the deaerator stream to the *boiler feed water (BFW) heater (Item E104)*. The temperature of this stream is set at about 92°C, and this will be used in the calculation of the stream specific enthalpy. We can calculate the liquid-phase, quench stream, enthalpy as:

> Vapour-Phase Enthalpy at 92°C – the Enthalpy of Vaporisation (Latent Heat) at the same temperature.

Using the appropriate polynomial from Appendix 3, we obtain a value for the vapour-phase enthalpy at 92°C as $-2.2700E + 05$ kJ/kmol.

Using steam tables, we obtain a value for the enthalpy of vaporisation (latent heat) at 92°C as 40986 kJ/kmol. Thus, the specific enthalpy of the quench stream at 92°C is:

$$-2.2700E + 05 - 40986 = -2.6799E + 05 \text{ kJ/kmol}$$

We have the total energy for the stream exiting the *converted gas boiler B (Item E103B)* and entering the *quench cooler as* $-7.8229E + 07$ kJ; to this, we now add the energy of the entering water stream in the quench:

$$= 50.0 \times -2.6799E + 05 = -1.3400E + 07 \text{ kJ}$$

If we add the enthalpies of the water streams entering the *quench cooler*, then we get a value of enthalpy as:

$-91629E + 07$kJ which is obtained by adding $\left(-7.8229E + 07 + \left(-1.3400E + 07\right)\right)$

We can use this energy figure to interpolate a value for the exit temperature.

Typically, if we take a value of exit temperature as 200°C, a table (Table 2.7) can be drawn up:

TABLE 2.7
Output Masses and Enthalpies for Quench Cooler

Component	Amount (kmol)	Specific Enthalpy (kJ/kmol)	Enthalpy
H_2	321.30	1.3760E+04	4.4211E+06
CO	12.28	−9.9962E+04	−1.2275E+06
CO_2	85.33	−3.7722E+05	−3.2189E+07
CH_4	0.24	−4.9897E+04	−1.1975E+04
N_2	112.59	1.3994E+04	1.5756E+06
H_2O	288.51	−2.2338E+05	−6.4447E+07
	820.25		−9.1878E+07

Taking the total enthalpy figure, a value of temperature can be fixed by comparing this value to the input enthalpy to the quench. With an adjustment of temperature to 201°C, the energy balance is obtained and the quench exit temperature is fixed at 201°C. This stream is the input stream to the *low-temperature shift converter.*

SUMMARY OF IMPORTANT PARAMETERS CALCULATED FOR CONVERTED GAS BOILER AND QUENCH COOLER

The energy available from converted gas boiler A can be calculated from an energy balance as −1.8523E+06 kJ. Energy available from converted gas boiler B can be calculated from an energy balance as −2.5879E+06 kJ. From both gas boilers, the total steam output can be calculated as 2852 kg.

The inlet water stream to the quench cooler can be calculated as 238.36 kmol.

Using the stated data of water uptake in the quench cooler output stream, the output from the quench cooler is calculated as 288.36 kmol.

Using the total stream enthalpy of the output stream from the quench cooler interpolation fixed the temperature of the stream as 201°C.

2.7 LOW-TEMPERATURE SHIFT CONVERTER (ITEM V103 ON THE BLOCK DIAGRAM)

REQUIRED

Calculation of equilibrium constant for the water–gas shift reaction:

$$CO + H_2O \leftrightarrow CO_2 + H_2$$

Equation used:

$$\ln K_p = \frac{4510.23}{T} - 4.196$$

Calculation of fractional conversion (z) for the carbon monoxide from the equilibrium constant expression.

Calculation of the component enthalpies for the outlet stream from the low-temperature shift converter.

2.8 MASS AND ENERGY BALANCE CALCULATIONS FOR THE LOW-TEMPERATURE SHIFT CONVERTER

The process stream from the *quench cooler* still contains a significant quantity of carbon monoxide. The temperature conditions in the high-temperature shift converter are essentially dictated by kinetic requirements to favour a fast reaction, and the high temperature introduces a limiting thermodynamic condition. The removal of the bulk of the remaining carbon monoxide is thermodynamically favoured, for the exothermic reaction, by a low temperature, but this means that the reaction kinetics are now a limiting factor. The low-temperature shift catalyst is important and expensive, but catalysts do exist based on zinc/copper/aluminium mixtures that can produce low carbon monoxide concentrations and maximise the hydrogen production. Such catalysts also have a fairly small approach to equilibrium.

The exit temperature from the *quench cooler* of 201°C is a low one for the catalyst used in the high-temperature shift reactor but, as has been pointed out, this low temperature is favourable for the thermodynamics within the reactor.

The inlet conditions for the low-temperature shift converter are set as follows:

	Inlet	Outlet
Temperature (°C)	201	220
Pressure (bar)	23.7	23.2

We have stated previously that the water–gas shift reaction to be considered is:

$$CO + H_2O \leftrightarrow CO_2 + H_2$$

And the K_p for the reaction can be calculated from:

$$\ln K_p = \frac{4510.23}{T} - 4.196$$

The entering stream is the exit stream from the *quench cooler* (Table 2.8).

TABLE 2.8

Component Masses for Input to Low-Temperature Shift Converter

Component	Amount (kmol)
H_2	321.30
CO	12.28
CO_2	85.33
CH_4	0.24
N_2	112.59
H_2O	288.51
	820.25

Note: Temperature is 201°C.

At a temperature of 220°C (474 K), K_P for this reaction, with an approach to equilibrium reported as 0°C, can be calculated as:

$$K_P = 141.5$$

Using the same approach to finding the equilibrium components as was used for this reaction in the *high-temperature shift converter*, we have to solve for z using the equation:

$$K_p = 141.5 = \frac{(85.33 + 12.28z)(321.3 + 12.28z)}{(12.28 - 12.28z)(288.51 - 12.28z)}$$

A solution for this equation gives a value of z as 0.933, and we find the following component quantities:

TABLE 2.9

Component Masses in Output Stream from the Low-Temperature Shift Converter

Component	Amount (kmol)
H_2	332.76
CO	0.822
CO_2	96.79
CH_4	0.24
N_2	112.59
H_2O	277.05
	820.25

Note: Temperature is 220°C. The CO quantity is now < 1 kmol.

For subsequent processing, it is necessary to find the energy content of the outlet stream based on the base enthalpies we have defined using the appropriate polynomial in Appendix 3.

TABLE 2.10

Output Masses and Enthalpies from Low-Temperature Shift Converter

Component	Amount (kmol)	Specific Enthalpy (kJ/kmol)	Enthalpy (kJ)
H_2	332.76	1.4354E+04	4.7764E+06
CO	0.822	−9.9353E+04	−8.1668E+04
CO_2	96.79	−3.7840E+05	−3.6625E+07
CH_4	0.24	−4.8988E+04	−1.1757E+04
N_2	112.59	1.4605E+04	1.6444E+06
H_2O	277.05	−2.2270E+05	−6.1699E+07
	820.25		−9.1997E+07

Note: Temperature is 220°C.

The process gas stream is now low in carbon monoxide, but the trace amounts are still present and there is still significant carbon dioxide present. Before the synthesis gas passes to the ammonia production stage, the carbon oxides are removed using a methanation process. Methanation uses reactions that have already been encountered at the primary reformer stage.

Essentially, the methanation uses a reverse methane–steam reaction:

$$CO + 3H_2 \leftrightarrow CH_4 + H_2O$$

If this is combined with the water–gas shift reaction, we can write an equation:

$$CO_2 + 4H_2 \leftrightarrow CH_4 + 2H_2O$$

The carbon oxides are removed with the formation of methane.

In terms of energy integration of the plant, there are a number of energy transfer steps that are carried out before the absorption and methanation steps.

In terms of the plant services after the low-temperature shift converter, after the exothermic reaction has occurred, the process gas stream energy can be used in a heater step to adjust the temperature of a water stream entering the boiler; this is Item E104, the high-pressure BFW heater.

SUMMARY OF IMPORTANT PARAMETERS CALCULATED FOR LOW-TEMPERATURE SHIFT CONVERTER

Carbon monoxide was removed in the low-temperature shift converter.

At a temperature of 220°C (474 K) and with an approach to equilibrium given as 0°C, K_p was calculated as 141.5.

Using the equation for K_p in terms of partial pressures, a value for the fractional conversion of carbon monoxide was calculated as 0.933.

The component masses are calculated and reported in Table 2.9.

The enthalpy of the exit stream at 220°C is calculated for subsequent input to the high-pressure BFW heater.

2.9   HIGH-PRESSURE BFW HEATER, ITEM E104

REQUIRED

A check as to whether water will condense from the syngas stream when this stream gives up energy to the BFW heater.

Calculation of amount of water to be condensed from the syngas stream as energy is transferred.

Calculation of the total energy to be transferred to the BFW heater.

Calculate the amount of steam condensing in the steam drum.

In considering the BFW, there has to be an acknowledgement that there may be a phase change involved in the heating of the BFW using the process gas exiting the low-temperature shift converter. One of the first calculations that will be involved is determining a *dew point*.

If we take the exit conditions from the converter, we assume, realistically, that there is a 3°C drop in the temperature of the process gas stream; thus, we know:

	Inlet	Outlet
Temperature (°C)	217	165
Pressure (bar)	23.2	22.7

The water to be heated will, initially, be considered to be at 90°C and will be processed at a pressure of 63.8 bar.

2.10   ENERGY BALANCE FOR BFW HEATER (INCLUDING PHASE CHANGE CHECKS)

Initially, the dew point of the process gas has to be determined related to any phase change. The water partial pressure in the process stream can be determined by first fixing the water mole fraction in the stream. Using the figures in Table 2.10, the water mole fraction is the moles water/total stream moles $= 277.05/820.25 = 0.338$. The stream has a total pressure of 22.7 bar, and hence using the equation:

$$p_{H_2O} = y_{H_2O}P$$

where p_{H_2O} is the water partial pressure, y_{H_2O} is the water mole fraction in the stream and P is the total pressure:

$$\text{Thus, } p_{H_2O} = 0.338 \times 22.7 = 7.67 \, \text{bar.}$$

Using steam tables, we can interpolate a temperature corresponding to this pressure as 168°C. We can now do a simplified calculation to fix the energy given up by the syngas stream. From experience, any condensation that may occur at or near the dew point does so over a range of temperatures. For the purposes of our calculation, we will assume (for simplicity) that we will cool to 165°C. On that basis, we will consider the cooling of the process gas from 217°C to 165°C as it gives up energy to the high-pressure BFW heater.

We can draw up the following table where the ΔH term has been calculated from the polynomial expression in Appendix 3 for each component and represents the difference in the input and output specific enthalpies:

TABLE 2.11

Masses and Output Enthalpies for Process Stream exiting HPBFWH

Component	kmol	ΔH (kJ/kmol)	Enthalpy (kJ)
H_2	332.76	1538	5.1179E+05
CO	0.822	1572	1.2920E+03
CO_2	96.79	2147	2.0781E+05
CH_4	0.24	2300	5.5200E+02
N_2	112.59	1590	1.7902E+05
H_2O	277.05	1759	4.8733E+05
Total	820.25		1.3878E+06

Taking the temperature as 165°C, we note that the corresponding stream pressure is 22.7 bar as given in the initial condition table.

At 165°C, using steam tables the corresponding saturation pressure for the water is 7 bar. Thus, in a total stream pressure of 22.7 bar, the partial pressure of the water is 7 bar and the partial pressure of the remaining dry gas must be $22.7 - 7.0 = 15.7$ bar.

On this basis, the ratio of water to dry gas is $7/15.7 = 0.466$.

From Table 2.11, the dry gas amount can be calculated as 543.2 kmol by simply subtracting the mass of water from the total stream mass.

It is realistic to assume that some water does condense at 165°C. In relation to this, we can calculate the amount of water that would be left in the process gas; this amount must be:

$543.2 \times 0.466 = 253.13$ kmol, i.e. amount of dry gas multiplied by the ratio of water to dry gas.

The amount of condensate must be:

$277.05 - 253.13 = 23.92$ kmol,

i.e. the amount of water initially coming through in the process stream (Table 2.11) minus the amount of water now coming through in the process stream.

Knowing the amount of condensate, we can calculate the amount of heat given up corresponding to this amount of water vapour condensing.

From steam tables, we can find that at 165°C, the enthalpy of vaporisation is 2067 kJ/kg; thus, making the necessary unit adjustment (kmol to kg), we can calculate the heat given up as:

$$23.92 \times 2067 \times 18 = 8.8997E + 05\,kJ$$

Taking the energy figure from Table 2.11, we can calculate a total energy exchange of:

$$8.8997E + 05 + 1.3878E + 06 = 2.2778E + 06\,kJ.$$

We can now calculate the temperature rise of the BFW.

On the basis of the process streams used in the production of the synthesis gas, the amount of BFW to be processed has been found to be 12745 kg. If we calculate that 99% of the energy available goes into this water, then the enthalpy per kg of BFW can be calculated as:

$$2.2778E + 06 \times 0.99/12745 = 176.9\,kJ/kg$$

As previously stated, the water is initially at a temperature of 90°C; from steam tables, it is possible to find that water specific enthalpy at this temperature is 376.9 kJ/kg.

Using the figure just calculated, we can write that the specific enthalpy at output must be:

$$176.9 + 376.9 = 553.8\ kJ/kg.$$

Using steam tables, this corresponds to an interpolated temperature of 132.3°C.

We can now calculate the heat needed to raise the stated amount of BFW to its boiling point taking the original specification that the water would be processed at a pressure of 63.8 bar.

At this pressure, water specific enthalpy is 1225 kJ/kg; thus, the heat required is:

$$12745 \times (1225 - 553.8) = 8.5544E + 06\ kJ.$$

From steam tables, if we interpolate a value for the enthalpy of vaporisation (latent heat) for water at 63.8 bar, we find a value of 1557 kJ/kg. Hence, the amount of steam condensing in the steam drum will be $8.5544E + 06/1557 = 5494\,kg$.

SUMMARY OF IMPORTANT PARAMETERS
CALCULATED FOR HIGH-PRESSURE BFW HEATER

The amount of water condensed from the syngas stream as energy is transferred from this to the BFW is 23.92 kmol.

The total energy exchanged is calculated as 2.2778E + 06 kJ.

The exit temperature of the BFW is 132.3°C.

The amount of water condensing in the steam drum is 5494 kg.

REFERENCE

1. 'Steam Tables' G F C Rogers, Y R Mayhew, *Thermodynamic and Transport Properties of Fluids*, 5th Edition, Blackwell Publishing, Oxford, UK.

3 Carbon Dioxide Removal and Stream Energy Interchange

3.1 CARBON DIOXIDE REBOILER AND KNOCKOUT POT (ITEMS E105 AND V104 ON THE BLOCK DIAGRAM)

REQUIRED

Summary and description of the energy input in the carbon dioxide reboiler and a description of the system for using the reboiler and regenerator in the basic Benfield solvent system.

3.2 DESCRIPTION OF CARBON DIOXIDE REMOVAL USING THE BENFIELD PROCESS

Energy has now been recovered from the process gas stream and has been calculated. This energy is available through the boiler feed water (BFW). One of the vital steps on the production of the syngas stream is now the removal of the carbon dioxide from this stream. This has been investigated over the time that ammonia gas has been produced by the process. It was pointed out earlier that the carbon dioxide can be, and sometimes is, removed by a process known as pressure swing adsorption (PSA). Essentially, this process requires the use of an effective solid to adsorb the carbon dioxide gas from the process stream. By control of the operating pressure, the adsorption takes place at this controlled pressure and desorption takes place as the pressure 'swings' to a lower controlled pressure. The process can take place at relatively low temperatures. There are some disadvantages: water carried over in the process stream can grossly affect the adsorption. Some of the high-pressure adsorption does have to take place at higher pressures which has vessel and economic implications.

Historically, the carbon dioxide has been removed by gas absorption. The solvent used in the absorption process is the key to the carbon dioxide recovery. Many plants employ mixtures of amines that have a proven record in removing the carbon dioxide. Many carbon dioxide removal and recovery absorptions used a solvent known as Benfield in the Benfield recovery process [1]. The Benfield solvent is essentially a hot potassium carbonate solution that absorbs the carbon dioxide in the process stream. The Benfield process consists of an absorption step and a regeneration step.

Basic Benfield Process Schematic.

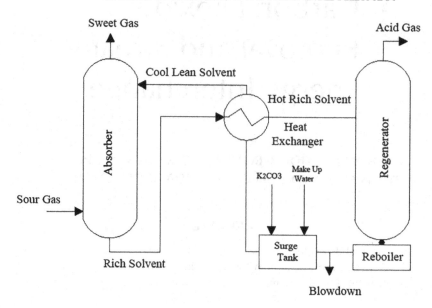

In classical gas absorption, the gas feed enters a packed column at the bottom of the column. The Benfield solution enters at the top of the column. The carbon dioxide is absorbed into the solvent, and hence, the process stream exits the column at the top with the carbon dioxide removed. As in any sensible process, the solvent then has to be 'regenerated'; i.e., the carbon dioxide has to be removed from the solvent and the stripped solvent recycled back to the absorption stage. This step requires a carbon dioxide reboiler to strip the gas out of the solvent, and it is this stage that shows on the block diagram as E105. Essentially, the energy balance required for E105 is very similar to that carried out for E104, the high-pressure BFW heater (BFWH), so only the final input and output stream quantities and conditions are recorded.

The stream exiting E105 enters V104, a knockout pot. This vessel separates condensate from the converted gas stream so that liquid water does not enter the absorber. The condensate is acidic because of the presence of dissolved carbon dioxide. It could be used for recycle to the boiler after suitable treatment.

The detailed mass and energy balances continue by consideration of the carbon dioxide using Benfield.

SUMMARY OF IMPORTANT PARAMETERS
CALCULATED FOR ABSORPTION PROCESS

No parameters are explicitly laid out in this section. An explanation of the Benfield process for absorbing the carbon dioxide is given, and the detailed output figures will be used in subsequent operations.

3.3 ABSORBER (ITEM V105 ON THE BLOCK DIAGRAM)

REQUIRED

The amount of carbon dioxide removed in the absorber. This is calculated on the basis of a known removal potential for the solvent.

Adjustment for small hydrogen loss.

Calculate the amount of water vapour that will exist in each of the exit streams from the total absorption system.

Calculate the exit temperature of the exit acid gas stream.

The process stream, having given energy for use in the carbon dioxide regenerator reboiler, now enters the absorber for removal of carbon dioxide prior to feeding on to the ammonia synthesis unit. Carbon dioxide is a serious catalyst poison and has to be removed.

For the purposes of the mass and energy balance, the amount of carbon dioxide to be removed has to be specified as a working operational parameter. The process stream emerges from the knockout pot having had reboiler energy removed and now at a temperature of 120°C.

The current stream conditions can be summarised as follows:

	Inlet	Outlet
	Converted Gas	Stripped Gas
Temperature (°C)	120	40
Pressure (bar)	22.6	22.0

3.4 MASS BALANCE CALCULATION FOR ABSORBER

The carbon dioxide removed depends on the efficiency of the absorption, but for the purposes of the mass balance, it is assumed that the carbon dioxide is removed to 0.15% on a dry basis. During the absorption, there will be a small amount of hydrogen gas lost. This is a processing issue, but for the purpose of this calculation, the loss will be assumed to be equal to 1% of the carbon dioxide removed.

Taking the mass figures exiting the BFWH (boiler feed water heater) which have not changed, we can write the new mass balance for the absorber:

With reference to the schematic, it will be seen that the mass and energy balance is over three streams. Essentially, the process gas (sour gas in the schematic) to be treated is fed to the bottom of the absorber and the carbon dioxide is absorbed by the Benfield solvent that is supplied at the top of the absorption column (cool, lean solvent in the schematic) and produces counter-current flow contact. The stream from which carbon dioxide has been removed is the stripped gas stream (sweet gas in the schematic). The solvent now has to be regenerated for further use. This requires the solvent,

TABLE 3.1

Mass Balance Figures for Absorber

Component	Inlet (kmol)	Stripped Gas Output (kmol)	Acid Gas Output (kmol)
H_2	332.76	332.76-0.96 = 331.8	0.96
CO	0.822	0.822	
CO_2	96.79	543.2 × 0.0015 = 0.815	95.98
CH_4	0.24	0.24	
N_2	112.59	112.5	
H_2O			
Total	543.2 (dry basis)	446.17	96.94

with absorbed carbon dioxide, to be fed to a regeneration column employing a carbon dioxide reboiler where the carbon dioxide is removed from the solvent and taken off as an acid gas stream (acid gas in the schematic). The solvent is then returned to the absorption column as an internal recycle stream.

Given the nature of the process described, water vapour will enter the outlet streams. The amounts can be calculated:

The acid gas stream is assumed to exist at 90°C as it exits the absorber and its pressure is slightly above the atmospheric pressure.

From steam tables, at 90°C, the component water vapour pressure can be found as 0.701 bar. Thus, for a total stream pressure of 1.2 bar the partial pressure of the remaining gas is taken as $1.2 - 0.701 = 0.499$ bar.

Thus, the steam-to-gas ratio can be calculated as:

$$0.701/0.499 = 1.405.$$

Table 3.1 shows the acid gas output as 96.94 kmol; thus, in this stream, the amount of water vapour will be:

$$96.94 \times 1.405 = 136.2 \, kmol.$$

In the same way as for the stripped gas stream, this will ultimately exist at 40°C and 22.0 bar.

From steam tables, we can find at 40°C that the component vapour pressure is 0.0738 bar.

As before, we can take the partial pressure of the remaining gas stream as:

$$22.0 - 0.0738 = 21.93 \, bar.$$

The steam-to-gas ratio can now be calculated as:

$$0.0738/21.93 = 0.0034.$$

In this stream, the amount of water vapour can now be calculated as:

$$446.17 \times 0.0034 = 1.517 \, \text{kmol}$$

where 446.17 kmol is the total mass of the dry stripped gas output stream. 1.517 kmol is the quantity of water in the stripped gas stream exiting the absorber.

Using the figures calculated for the water vapour outputs, the total water leaving in the absorber streams can be calculated as:

$$136.2 + 1.517 = 137.72 \, \text{kmol}.$$

The unshown calculation on the CO$_2$ reboiler that used energy from the process stream to regenerate the solvent showed that an amount of water, 52.47 kmol, was left in the process stream after the operation. The mass difference between the total amount of water leaving the absorber, 137.72 kmol, and the amount of water leaving the CO$_2$ regenerator, 52.47 kmol, is made up by cooling the acid gas stream and returning the condensate to the system.

To do this cooling calculation, we acknowledge that we have already accounted for 1.517 kmol of this stream. We consider:

$$52.47 - 1.517 = 50.95 \, \text{kmol}$$

With reference to Table 3.1, it can be seen that on a water-free basis, the total stream kmols are 96.94.

The mole fraction of water is then:

$$\frac{50.95}{50.95 + 96.94} = 0.345$$

The partial pressure of the water can then be calculated using the equation $p_i = y_i P$. We know $P = 1.2$ bar.

The partial pressure of water is then:

$$p_i = 0.345 \times 1.2 = 0.414 \, \text{bar}$$

Using steam tables, we can interpolate a value for stream temperature as 76.7°C. This is the temperature that the acid gas stream is cooled to.

SUMMARY OF IMPORTANT PARAMETERS CALCULATED FOR ABSORBER

The amount of carbon dioxide remaining after absorption is 543.2 × 0.0015 = 0.815 kmol.

The amount of water vapour initially leaving in the exit streams is calculated as 137.72 kmol.

The amount of water leaving in the stripped gas exit stream is 1.517 kmol.
The amount of water leaving the regenerator is 52.47 kmol.

The difference between the total water amount leaving and the amount of water leaving the regenerator is made up by condensing the acid gas stream to a calculated temperature of 76.7°C.

3.5 ENERGY INTERCHANGER (ITEM E106 ON THE BLOCK DIAGRAM)

REQUIRED

Energy transfer between four process streams:

Stream (1): stripped gas stream from absorber
Stream (2): feed to methanator (Item V107)
Stream (3): exit stream from reformer gas boiler
Stream (4): feed to high-temperature shift converter 1

The stream energies were fixed or had been calculated.
The final energy balance has to be carried out.

The energy released and absorbed in process operations has always been a feature of the operations. It was apparent that there was a potential for energy wastage simply because the energy released in a particular operation was often transferred to the surroundings rather than seeing if there was a physically sensible, and possible, opportunity for the redistribution of the energy within the process.

In the 1970s, pinch technology was developed which allowed optimisation of energy stream transfer. In this text, energy transfer is allowed to take place between two of the process streams in an interchanger. There is no detailed analysis of process integration. With fixed or calculated stream temperatures, the energy transferred from one stream to another is calculated using the techniques employed in the previous calculations.

The two process streams involved in the energy interchange are the exit stream from the reformed gas boiler and the exit stream from the absorption column.

3.6 ENERGY BALANCE FOR ENERGY INTERCHANGER

AS has been discussed previously, in most of the enthalpy calculations carried out, the values of enthalpy calculated have been stated in Excel numerical form and rounded off. The enthalpy figures for the interchanger where four streams are involved have been recorded in scientific, exponential form for the convenience of presentation. The calculations are still valid, and the results are reported in the summary.

For the purposes of the calculation, the following conditions have been fixed or calculated (Table 3.2):

TABLE 3.2
Conditions of Input Streams to Absorber

	Reformed Gas Boiler Stream at Interchanger		Absorber Stripped Gas Stream at Interchanger	
	Inlet	Outlet	Inlet	Outlet
Temperature (°C)	535	389	40	325
Pressure (bar)	25.5	25.3	22.0	21.8

The stripped gas stream exiting from the absorber into the interchanger becomes the input stream to the methanator at a temperature fixed at 325°C (Table 3.3).

TABLE 3.3
Masses and Component Enthalpies for Streams Entering and Leaving the Interchanger

Component	Stream (kmol)		Stream Component Enthalpy (kJ)			
	1,2	3,4	1	2	3	4
H$_2$	331.8	269.17	3.0167E+06	5.8095E+06	6.4620E+06	5.2389E+06
CO	0.822	64.41	−8.6040E+04	−7.8977E+04	−5.7411E+06	−6.0786E+06
CO$_2$	0.815	33.2	−3.1233E+05	−3.0289E+05	−1.1984E+07	−1.2236E+07
CH$_4$	0.24	0.24	−1.3492E+04	−1.0514E+04	−7.5810E+03	−9.6810E+03
N$_2$	112.5	112.5	1.0241E+06	2.0041E+06	2.7239E+06	2.2253E+06
H$_2$O	1.504	290.45	−3.4398E+05	−3.2952E+05	−6.1487E+07	−6.2988E+07

Stream (1): stripped gas stream from absorber
Stream (2): feed to methanator (Item V107)
Stream (3): exit stream from reformed gas boiler
Stream (4): feed to high-temperature shift converter 1

This type of stream balance could fix the temperature of Stream 2. In this case, a figure of 325°C was fixed using operating data and the stream energy balance fixed; the results are given in Table 3.4.

In the energy balances carried out, it was stated at the start that each energy calculated would be calculated to a number of significant figures before the decimal point. For the convenience of presentation, the figures here are quoted in Excel scientific form.

TABLE 3.4

Energy Balance for Streams Entering and Leaving the Interchanger

Component	Enthalpy Stream 2–Enthalpy Stream 1	Enthalpy Stream 4–Enthalpy Stream 3
H_2	2.7928E+06	−1.2231E+06
CO	7.0630E+03	−3.3750E+05
CO_2	9.4400E+03	−2.5200E+05
CH_4	2.9780E+03	−2.1000E+03
N_2	9.8000E+05	−4.9860E+05
H_2O	1.4460E+04	−1.5010E+06
Total	3.8067E+06	−3.8143E+06

SUMMARY OF IMPORTANT PARAMETERS CALCULATED FOR INTERCHANGER

The stream temperatures were assumed or as previously calculated:

	Reformed Gas Boiler Stream at Interchanger		Stripped Gas Stream at Interchanger	
	Inlet	Outlet	Inlet	Outlet
Temperature (°C)	535	389	40	325

The stream enthalpies were calculated using the component polynomial expressions:

Enthalpy Stream 2–Enthalpy Stream 1	Enthalpy Stream 4–Enthalpy Stream 3
3.8067E+06	−3.8143E+06

The agreement is considered acceptable.

As with previous energy balances, these figures agree within <1% and the balance proceeds with Stream 2, the feed to the methanator (Item V107) going forward.

3.7 METHANATOR (ITEM V107 ON THE BLOCK DIAGRAM)

REQUIRED

Stoichiometric equations for removal of carbon oxides using the reverse methane–steam reaction and the reaction of carbon dioxide and hydrogen producing methane.

$$CO + 3H_2 \leftrightarrow CH_4 + H_2O \tag{3.1}$$

$$CO_2 + 4H_2 \leftrightarrow CH_4 + 2H_2O \tag{3.2}$$

Calculation of equilibrium constants for Reactions 1 and 2 introduced in chapter 1

The equilibrium constant for Reaction 2 can be calculated as:

$$K_{P2} = \frac{1}{K_{P1}.K_{P3}}$$

where K_{P3} refers to the equilibrium constant for water–gas shift reaction.

On the basis of the values of the equilibrium constants calculated, the values of z_1 and z_2, the fractional conversions of Reactions 1 and 2, can be calculated.

The component masses in the output stream can be calculated.

The N$_2$/H$_2$ ratio can be calculated.

The input and output stream enthalpies to be calculated and the balance checked.

We consider this operation with the following input and output conditions:

	Inlet	Outlet
Temperature (°C)	325	355
Pressure (bar)	21.7	21.3

In the production of ammonia syngas, the gas entering the ammonia synthesis loop has to be close to the stoichiometric ratio of nitrogen/hydrogen of 3:1. The gas entering the loop is processed to the desired product, ammonia, using a treated iron catalyst. This catalyst is particularly sensitive to the poisoning effect of carbon oxides, and efforts have to be made to reduce the amount of carbon monoxide and carbon dioxide in the syngas stream.

3.8   MASS BALANCE CALCULATION FOR METHANATOR

An inspection of Table 3.1 will show that the two oxides are present at a level of <1%. The methanator effectively reduces this level to almost non-detectable levels. The 'downside' of this process is that it requires hydrogen as a reactant, and this is taken from the syngas stream.

Essentially, the methanation process uses the reverse reactions of those considered to operate in the primary reformer.

For the purpose of the mass and energy balances in the methanator, the two equations considered are those previously discussed:

$$CO + 3H_2 \leftrightarrow CH_4 + H_2O \tag{3.1}$$

$$CO_2 + 4H_2 \leftrightarrow CH_4 + 2H_2O \tag{3.2}$$

Both equations allow the formation of methane in the final equilibrium mixture.

The first equation will be recognised as the reverse methane–steam reaction, and in the primary reformer, the second equation was taken as the water–gas shift reaction. There is a slight challenge at this point to demonstrate that the reversed water–gas shift reaction is not an independent equation from those to be used. We could always try to subtract equation 3.3 from equation 3.4 and see what appears. The K_P values for the methane–steam reaction and the water–gas shift reaction are given in Appendix 1. Basically, because we have the data for the methane–steam and water–gas shift reaction, it would be easier and technically acceptable to use these two reactions for calculations on the methanator.

If it was necessary to express a K_P value for Reaction 3.2 above, it can be demonstrated that K_P for that reaction, K_{P2}, can be written as:

$$K_{P2} = \frac{1}{K_{P1}, K_{P3}}$$

where K_{P1} refers to the equilibrium constant for the methane–steam reaction and K_{P3} refers to the equilibrium constant for the water–gas shift reaction:

which is quoted in Appendix 1 and appears here as equation 3.3

$$CO + H_2O \rightarrow CO_2 + H_2 \tag{3.3}$$

The reader might find it a useful exercise to confirm the quoted expression for K_{P2}

If we use Appendix 1 data, we can find that K_P for the methane–steam reaction at 355°C is 2.924×10^{-6} and K_P for the water–gas shift reaction is 16.811. On that basis, the value of K_{P2} is:

$$1/(K_{P1} \cdot K_{P3}) = 1/(2.924 \times 10^{-6} \times 16.811) = 20,344. \tag{3.4}$$

The value of K_P for the reverse methane–steam reaction (Equation 3.1 above) can be calculated as:

$$1/K_{P1} = 1/2.924 \times 10^{-6} = 341997.$$

Basically, with such large values of the K_p, both reactions are considered to go to completion at equilibrium.

The absorbed gas stream that enters the methanator has the following mass figures:

TABLE 3.5

Component Masses Entering the Methanator

Component	Stripped Gas Stream Input (kmol)
H$_2$	331.8
CO	0.822
CO$_2$	0.815
CH$_4$	0.24
N$_2$	112.5
H$_2$O	1.504
Total	447.68

If the two reactions considered to represent the methanation go to completion, then the output stream from the methanator will be as follows:

TABLE 3.6

Component Masses Entering and Leaving the Methanator

Component	Stripped Gas Stream Input (kmol)	Stripped Gas Stream Output from Methanator (kmol)
H$_2$	331.8	$331.8 - 0.822 \times 3 - 0.815 \times 4 = 326.07$
CO	0.822	0
CO$_2$	0.815	0
CH$_4$	0.24	$0.24 + 0.822 + 0.815 = 1.88$
N$_2$	112.5	112.5
H$_2$O	1.504	$1.504 + 0.822 + 0.815 \times 2 = 3.96$
Total	447.68	444.41

If the H$_2$/N$_2$ ratio is checked, it is found to be $331.8/112.5 = 2.9$.

The original aim was to produce the H$_2$/N$_2$ ratio as 3.0. Thus, at the end of the synthesis gas production, we would try to ensure that the CO content was negligible; the implication of this is that at the exit from the secondary reformer, the ratio $(CO + H_2)/N_2$ had to be 3/1. This enabled us to set up balances that were used to calculate the amounts of each component in the secondary reformer exit. The subsequent processing based on the data available indicated that the methanation stage was reducing the amount of hydrogen to an extent that would probably require some make-up hydrogen in the ammonia synthesis loop.

3.9 ENERGY BALANCE FOR METHANATOR

The component enthalpies can be calculated for each component at the stated stream temperatures using the polynomials for enthalpies as given in Appendix 3.

Input Stream Temperature = 325°C

TABLE 3.7
Component and Stream Enthalpies for the Input Stream to Methanator

Component	kmol	Input Enthalpy (kJ/kmol)	Stream Enthalpy (kJ)
H_2	331.8	1.7509E+04	5.8095E+06
CO	0.822	−9.6079E+04	−7.8977E+04
CO_2	0.815	−3.7165E+05	−3.0289E+05
CH_4	0.24	−4.3809E+04	−1.0514E+04
N_2	112.5	1.7814E+04	2.0041E+06
H_2O	1.504	−2.1909E+05	−3.2952E+05
		Sum	7.0917E+06

Output Stream Temperature = 355°C

TABLE 3.8
Component and Stream Enthalpies for the Output Stream from Methanator

Component	kmol	Output Enthalpy (kJ/kmol)	Stream Enthalpy
H_2	326.07	1.8422E+04	6.0069E+06
CO	0	0.0000E+00	0.0000E+00
CO_2	0	0.0000E+00	0.0000E+00
CH_4	1.88	−4.2210E+04	−7.9355E+04
N_2	112.5	1.8729E+04	2.1070E+06
H_2O	3.96	−2.1805E+05	−8.6349E+05
			7.1710E+06

The balance is within 1.2% and considered acceptable for this calculation.

SUMMARY OF IMPORTANT PARAMETERS
CALCULATED FOR THE METHANATOR

The equilibrium constants for Reactions 1 and 2 can be calculated as:

$$1/K_{P1} = 1/2.924 \times 10^{-6} = 341997.$$

$$K_{P2} = 1/(K_{P1} \cdot K_{P3}) = 1/(2.924 \times 10^{-6} \times 16.811) = 20,344.$$

With such high values of the equilibrium constants, the reactions can be considered to achieve complete conversion in the presence of the appropriate catalyst.

The values of z_1 and z_2 will be considered to be equal to 1.

The output component masses can be calculated as set out in Table 3.5.

The input and output component enthalpies can be calculated using the appropriate polynomial equations. The input and output energies can be calculated and the balance checked (Tables 3.6 and 3.7).

3.10 LOW-PRESSURE BFWH (ITEM E107 ON THE BLOCK DIAGRAM)

REQUIRED

Enthalpies for the input and output streams of the BFWH
    Checking the phase state of the water in the output stream after the cooling in the heat transfer.
    Calculating the energy available for heat transfer into the BFW.
    Calculating the amount of BFW that can be heated.

The stream leaving the methanator is at a relatively high temperature, and the energy available in the stream can be used to heat water for use in a low-pressure BFWH. The input and output stream conditions can be written as follows:

3.10.1 METHANATOR GAS

	Inlet	Outlet
Temperature (°C)	355	70
Pressure (bar)	21.3	21.1

3.10.2 BOILER FEED WATER

	Inlet	Outlet
Temperature (°C)	15	84

3.11 ENERGY BALANCE CALCULATION FOR LOW-PRESSURE BFWH

Table 3.8 gives the information on the input stream to the low-pressure BFWH, and the input and output pressure and temperature are quoted above. For the BFW, it is useful to use the stream energy from the methanator to raise the temperature to a recommended level between 80°C and 85°C. This means that less energy is required in the boiler itself and the preheating reduces the amount of bioactivity that might occur in the stream before its use in the boiler.

In this case, the output boiler temperature has been fixed at 84°C.

Knowing the output condition from the methanator allows the calculation of stream enthalpies using the polynomials given in Appendix 3.

The data are calculated as given in Table 3.8 and are as follows:

Component	Kmol	Enthalpy (kJ/kmol)	Stream Enthalpy
H_2	326.07	1.8422E+04	6.0069E+06
CO	0	0.0000E+00	0.0000E+00
CO_2	0	0.0000E+00	0.0000E+00
CH_4	1.88	−4.2210E+04	−7.9355E+04
N_2	112.5	1.8729E+04	2.1070E+06
H_2O	3.96	−2.1805E+05	−8.6349E+05
			7.1710E+06

Before the output stream at 70°C and the stream enthalpies can be calculated, in a similar fashion to the checking on previous streams where there was a possibility of water condensing from the stream, what will be the state (vapour or liquid) of the water in the stream?

From the figures given, the mole fraction of water in the stream is 3.96/444.41 = 0.009.

At a stream pressure of 21.1 bar, the partial pressure of the water is:

$$0.009 \times 21.1 = 0.189 \, bar.$$

Using steam tables, this fixes the saturation temperature for the water between 55°C and 60°C. This would indicate that the water in the stream will not condense at the conditions stated.

TABLE 3.9

Output Masses and Component and Stream Enthalpies for Low-Pressure Boiler Feed Water Heater

Component	kmol	Enthalpy (kJ/kmol)	Stream Enthalpy
H$_2$	326.07	9.9560E+03	3.2464E+06
CO	0	0.0000E+00	0.0000E+00
CO$_2$	0	0.0000E+00	0.0000E+00
CH$_4$	1.88	−5.5166E+04	−1.0371E+05
N$_2$	112.5	1.0019E+04	1.1271E+06
H$_2$O	3.96	−2.2773E+05	−9.0180E+05
Total	444.41		3.3680E+06

The energy available from the stream can then be calculated as the difference between the input and output stream enthalpies:

$$7.1710E + 06 - 3.3680E + 06 = 3.8030E + 06 \, kJ$$

Knowing that the BFW available is heated from 15°C to 84°C, we can calculate the amount of BFW taking a value of the specific heat capacity for water as 4.195 kJ/kg K:

$$3.8030E + 06 \, kJ / \left(4.195 \times (84 - 15)\right) = 13,138 \, kg$$

SUMMARY OF IMPORTANT PARAMETERS CALCULATED FOR LOW-PRESSURE BOILER FEED WATER HEATER (BFWH)

Inspection of the stream water partial pressure compared to the stream water saturation vapour pressure indicates that no water vapour in the stream will condense.

The stream enthalpies are calculated using the component enthalpies calculated using the polynomials. The values are reported in Tables 3.7 and 3.8.

The energy available for heating the BFW is the energy difference between input and output streams:

$$7.1710E + 06 - 3.3680E + 06 = 3.8030E + 06 \, kJ$$

Knowing the specific heat capacity of the water, the amount of BFW is:

$$3.8030E + 06 \, kJ / \left(4.195 \times (84 - 15)\right) = 13,138 \, kg$$

3.12   FINAL GAS COOLER (ITEM E108 ON THE BLOCK DIAGRAM)

REQUIRED

Total energy to be removed to cool the exit stream for the cooler calculated by finding the input and output stream enthalpies.

Calculation of the amount of water condensate produced in the syngas as the stream is cooled to 40°C.

Calculate the amount of energy associated with the production of the water condensate.

Using the total energy to be removed, calculate the amount of cooling water required in the final cooler.

The gas stream emerging from the BFWH has now to be finally cooled before being processed for the ammonia synthesis section. The final cooling temperature is specified as 40°C. To be able to carry out any mass or energy balancing on this stream requires properties of the stream to be specified:

Final cooler stream input and output conditions:

3.12.1   METHANATOR GAS

	Inlet (°C)	Outlet (°C)
Temperature (°C)	70	40
Pressure (bar)	21.1	20.9

3.12.2   COOLING WATER

	Inlet (°C)	Outlet (°C)
Temperature (°C)	15	25

3.13   ENERGY BALANCE CALCULATION FOR FINAL GAS COOLER

In cooling, the first requirement is to calculate the amount of sensible heat that has to be removed from the final gas stream:

The necessary figures can be tabulated:

The input stream can take the data from Table 3.9 reproduced here:

Component	kmol	Enthalpy (kJ/kmol)	Stream Enthalpy
H$_2$	326.07	9.9560E+03	3.2464E+06
CO	0	0.0000E+00	0.0000E+00
CO$_2$	0	0.0000E+00	0.0000E+00
CH$_4$	1.88	−5.5166E+04	−1.0371E+05
N$_2$	112.5	1.0019E+04	1.1271E+06
H$_2$O	3.96	−2.2773E+05	−9.0180E+05
Total	444.41		3.3680E+06

The output stream relates to a stream temperature of 40°C (Table 3.10):

TABLE 3.10
Output Masses and Component and Stream Enthalpies for Final Cooler

Component	kmol	Enthalpy (kJ/kmol)	Stream Enthalpy (kJ)
H$_2$	326.07	9.0920E+03	2.9646E+06
CO	0		0
CO$_2$	0		0
CH$_4$	1.88	−5.6217E+04	−1.0569E+05
N$_2$	112.5	9.1030E+03	1.0241E+06
H$_2$O	3.96	−2.2871E+05	−9.0570E+05
Total	444.41		2.9773E+06

Taking the enthalpy difference between the input and output streams:

$$2.9773E+06 - 3.3680E+06 = -3.9070E+05\,kJ \tag{3.5}$$

At a temperature of 40°C, a small amount of condensate is expected from the water element of the stream.

Using steam tables, we can find the saturation vapour pressure of water as 0.07375 bar. We know the final stream pressure is 20.9 bar.

Using these figures, the water vapour mole fraction is:

$$0.07375/20.9 = 0.0035$$

The water vapour amount must then be:

$$444.41 \times 0.0035 = 1.56\,kmol, \text{where } 444.41 \text{ is the total stream moles.} \tag{3.6}$$

The total water content of the stream is 3.96 kmol; thus, the amount of water condensing must be:

$$3.96 - 1.56 = 2.40 \text{ kmol of condensate. kg of condensate is then } 2.40 \times 18 = 43.20 \text{ kg.}$$
$$(3.7)$$

At 40°C, the enthalpy of vaporisation (latent heat) can be read from steam tables as 2406.2 kJ/kg.

The energy associated with the phase change is then:

$$43.20 \times 2406.2 = 1.0395E + 05 \text{ kJ} \tag{3.8}$$

The total heat load is then:

$$3.9070E + 05 \text{ kJ} + 1.0395E + 05 = 4.9465E + 05 \text{ kJ} \qquad (3.9) \text{ (see (3.5) and (3.8))}$$

Thus, the cooling water required for the gas stream can be calculated.

In going from 15°C to 25°C, the change in specific enthalpy can be found from steam tables:

$$104.8 - 62.9 = 41.9 \text{ kJ/kg} \tag{3.10}$$

The water needed for cooling will be:

$$4.9465E + 05/41.9 = 11,805 \text{ kg.} \tag{3.11}$$

SUMMARY OF IMPORTANT PARAMETERS
CALCULATED FOR FINAL HEATER

The difference between the input and output stream enthalpies is:

$$2.9773E + 06 - 3.3680E + 06 = -3.9070E + 05 \text{ kJ}$$

The amount of water still existing as water vapour in the output stream is:

$$1.56 \text{ kmol.}$$

Knowing the total amount of water exiting the cooler, the amount of condensate is calculated:

$$3.96 - 1.56 = 2.40 \text{ kmol}$$

The amount of energy associated with the water phase change is:

$$1.0395E + 05 \text{ kJ}$$

The total energy to be removed in the cooling is:

$$3.9070E + 05 \text{ kJ} + 1.0395E + 05 = 4.9465E + 05 \text{ kJ}$$

To handle this heat load, the total amount of water required is:

$$11,805 \text{ kg}$$

REFERENCE

1. Arthur L Kohl, Richard Nielson, 1997, *Gas Purification*, Elsevier, Netherlands.

4 Compression, Preheat and Desulphurisation

4.1 COMPRESSION, PREHEAT AND DESULPHURISATION OF THE NATURAL GAS STREAM (ITEMS C101, E101A AND E101B AND V101 ON THE BLOCK DIAGRAM)

Of these items the first considered is:

4.2 NATURAL GAS FEEDSTOCK COMPRESSOR (ITEM C101 ON THE BLOCK DIAGRAM)

REQUIRED

Basic equations for polytropic expansion.

Equations for $\gamma = \dfrac{C_P}{C_V}$ and calculation as required.

Calculation of the polytropic index.

Calculation of compression stages and exit temperature on exiting compression stages.

It is not part of the purpose of this book to investigate the equations and methods required to produce a working compressor for the process. It would probably be legitimate to say 'here is a compressor and the output pressure and temperature from the compressor are x and y and...'. In practice, a small explanatory note will be included, but it will be important to specify the output conditions of the process stream from the compressor before the stream is fed into the feed preheater.

It is proposed to feed the gas into the synthesis process at a pressure of 32.8 bar with the natural gas being available at a pressure of 5.1 bar. The classic approach to compression in thermodynamics is through the simple idea of a piston in a cylinder with an ideal gas being compressed as the piston is pushed into the cylinder, decreasing the volume and increasing the pressure. This idea can be incorporated into a reciprocating compressor.

The conditions under which the compression is carried out are classically investigated, initially under isothermal, isobaric and adiabatic conditions. Calculations are often carried out by assuming adiabatic conditions and then adjusting to the real conditions being used.

In assessing the compression, one of the basic parameters required is the ratio of the output pressure required to the input pressure actually fixed, $\dfrac{P_2}{P_1}$.

The ratio $\dfrac{P_2}{P_1}$ can be calculated as $32.8/5.1 = 6.43$.

For a high ratio such as this, the indications are that the compressor should be split into two stages.

Such a configuration would normally operate with a heat exchanger (intercooler) between the two stages. This intercooler normally is included to reduce the interstage steam temperature, which increases the overall efficiency of the compressor system and obviously changes the output temperature.

The ratio of the input and output pressures to the two stages should be as close to equality as can reasonably be achieved. The figures given in this section assume a pressure drop over the intercooler of 0.2 bar.

The pressures given on the block diagram indicate that the output pressure from the first stage is taken as 13.0 bar. With a pressure drop through the intercooler of 0.2 bar, the input pressure to the second stage is 12.8 bar. The second-stage output pressure has already been indicated as 32.8 bar. *As a check, the pressure ratio from the first stage is:*

$$\frac{P_2}{P_1} = \frac{13.0}{5.1} = 2.55$$

For the second stage, the pressure ratio is:

$$\frac{P_2}{P_1} = \frac{32.8}{12.8} = 2.56$$

which is acceptable.

For the feedstock being compressed, under the conditions described, the stream will not be acting ideally. The treatment and calculation will be based on a parameter called the polytropic index. The ideal efficiency for the compressor would include an 'ideal index' normally designated as γ where γ can be written as:

$$\gamma = \frac{C_P}{C_V}$$

where C_P is the specific heat capacity at constant pressure and C_V the specific heat capacity at constant volume.

For the polytropic case, we need the polytropic index, and a study of the polytropic compression will allow a calculation through the manufacturer-supplied efficiency of the compressor (η). Assume an efficiency of 0.7 has been supplied; then, we can write an expression in terms of the index n as:

$$\frac{n-1}{n} = \frac{\gamma-1}{\eta}$$

For the feedstock being considered, which is mainly methane, the value of γ can be found as 1.31 and the polytropic index fixed as:

$$\frac{n-1}{n} = \frac{1.31-1}{1.31 \times 0.7}$$

Hence, n can be calculated as 1.51.

If more detailed calculation in relation to the compressor was to be carried out, this polytropic index could be used to calculate power requirement (H) of the compressor:

$$H = \frac{nP_1}{n-1}\left[\left(\frac{P_2}{P_1}\right)^{\left(\frac{n-1}{n}\right)} - 1\right]$$

However, for the current mass and energy balances, the calculation of the exit temperature from the first compression stage can be calculated as:

$$\text{Temperature change} = T_{in}\left(\left[\frac{P_2}{P_1}\right]^{\frac{n-1}{n}} - 1\right)$$

where subscripts 1 and 2 refer to input and output pressures and n is the polytropic index.
  For the first compression stage:
  Substituting:

$$= 298\left[\left(\frac{13.0}{5.1}\right)^{\frac{1.51-1}{1.51}} - 1\right]$$

$$= 110.91$$

The outlet temperature is $110.91 + 298 = 408.91$ K $= 135.9°C$.
  After the intercooler between the two compression stages, the inlet temperature to the second stage is 40°C with a stream pressure of 32.8 bar. The output temperature from this stage can then be calculated as:

$$= 313\left[\left(\frac{32.8}{12.8}\right)^{\frac{1.51-1}{1.51}} - 1\right]$$

$$= 117.06$$

The outlet temperature is $313 + 117.06 = 430.06$ K $= 157.1°C$.
  This was recorded as 160°C for the purposes of feeding onto the feed preheater.

SUMMARY OF IMPORTANT PARAMETERS CALCULATED

The ratio of the inlet and outlet pressures, 32.8/5.1 = 6.43, **indicates that the compressor should** operate in two stages.
  The polytropic index n is calculated as 1.51.
  The outlet temperature from the first-stage compression is calculated as 135.9°C.
  The output temperature from the second-stage compressor is calculated as 157.1°C.

The outlet stream from the compressor into the preheater can now be considered.

4.3   FEEDSTOCK PREHEATER (ITEMS E101A AND E101B ON THE BLOCK DIAGRAM)

REQUIRED

Values of the specific heat capacity, C_P, at constant pressure as a function of T, the temperature for all the components in the natural gas feed stream.

Integration of the equations for C_P as a function of T.

Using the integrated equations, calculation of the component stream enthalpies for all the components in the feed stream.

Calculation of the stream enthalpies for the inlet and outlet streams.

Calculation of the energy requirements to heat the feed stream.

Use of the enthalpy polynomials from Appendix 3 to calculate the component stream enthalpies for all the components in the feed stream.

Calculation of the stream enthalpies for the inlet and outlet streams.

Calculation of the energy requirements to heat the feed stream.

Composition of the flue gas stream to be used to heat the natural gas feed stream.

Splitting of the feed stream heater into two stages to avoid temperature crossover.

Use of the enthalpy polynomials from Appendix 3 to calculate the component stream enthalpies for all the components in the flue gas stream.

Calculation of the stream enthalpies for the inlet and outlet streams based on supplied temperatures.

Calculation of the available energy from the two-stage heaters.

Balancing of the energy required and the energy available.

Realistically, if we consider a 5°C drop in temperature after the compression stage, the input conditions to the heater can be written as follows:

Natural Gas Feedstock	Input	Output
Temperature (°C)	155	410
Pressure (bar)	32.8	32.5

The preheater operates in two stages, and the energy required by the preheater is supplied by the energy from the flue gas. A range of energy requirements have been demonstrated to be satisfied by process stream heat exchange with other requirements. There is still a basic energy demand from the plant, and this is supplied by burning fuel. In this case, the fuel burned is natural gas of the same composition as the feedstock. A flue gas of known composition is used. An inspection of the chapter section dealing with the primary reformer will show that calculations were carried out on the burning of natural gas as a source of energy for the reformer. An amount of flue gas was produced. A proportion of this gas was used as an energy source on plant that is not described here. The flue gas amounts used here are quoted. It will be noted that the flue gas is still of the same composition as that produced in the burning of the natural gas for the primary reformer. This flue gas stream will also be used to carry out a series of calculations in Chapter 6. This will incorporate a series of operations requiring mass and energy balancing but with the suggested quantitative balances incorporated in a different location.

It will obviously be necessary to carry out energy balances on the two streams involved. At this point in the calculations, an opportunity will be taken to demonstrate a number of different calculation methods for the necessary enthalpy data. It is obviously important that the different methods point to the same answer.

The first method uses specific heat capacities at constant pressure C_p. The calculations are explained and carried out in Appendix 3. The values obtained through the use of the C_p polynomials are reported below.

4.4 CALCULATION OF PREHEATER MASS AND ENERGY BALANCE

The mass balance is already fixed. The preheat will be applied to the compressed natural gas stream. The important data involved in the preheater are those involved in the energy balancing. It is proposed to use energy from a flue gas stream to carry out the preheat. The flue gas is created by fuel burning on the plant to supply the plant overall energy requirements. Some details of the ancillary equipment involved in creating and handling the flue gas stream will be the subject of exercises scheduled in Chapter 6.

For the calculations carried out here, stream temperature will be taken from the plant block diagram and the flue gas temperature will be supplied from the separate calculations carried out on this stream.

The data supplied in Table 4.1 were used and the energy balances across the two preheaters carried out.

There were indications that for the preheating, the preheater would have to be split into two stages to prevent a temperature crossover as energy was transferred from the flue gas to the natural gas stream. On this basis, the calculations are carried out on preheater E101A and preheater E101B.

The input and output temperatures were fixed as follows:

Item	E101A		E101B	
	Input	Output	Input	Output
Temperature (°C)	155	328	328	410

The input temperature is fixed at 155°C because a 5°C temperature drop has been allowed on the stream exiting the compressor.

Using the integrated expressions for the C_p values of the natural gas, the flowing stream energy differences were computed:

TABLE 4.1
Enthalpy Calculations for Preheater E101A

Component	Initial Moles (kmol)	Input Component Enthalpy (kJ/kmol)	Input Stream Enthalpy (kJ)	Output Component Enthalpy (kJ/kmol)	Output Stream Enthalpy (kJ)
CH_4	94.68	1.3232E+04	1.2528E+06	2.1481E+04	2.0338E+06
C_2H_6	3.00	1.6888E+04	5.0664E+04	3.068E+04	9.2040E+04
C_3H_8	0.50	2.2372E+04	1.1186E+04	4.2352E+04	2.1176E+04
C_4H_{10}	0.40	2.9837E+04	1.1935E+04	5.4174E+04	2.1670E+04
N_2	1.40	1.2692E+04	1.7769E+04	1.7829E+04	2.4961E+04
H_2S	0.02	1.4339E+04	2.87E+02	2.083E+04	4.166E+02
Total			1.3446E+06		2.1941E+06

The details of the calculational method are given in Appendix 3. It should be noted that, provided the lower temperature set in the integration is 0 K, the results can be directly compared.

4.5 ENERGY VALUES CALCULATED USING C_p DATA

SUMMARY FOR PREHEATERS E101A AND E101B USING C_p DATA FOR ENTHALPY CALCULATIONS:

Item	E101A		E101B	
	Input	Output	Input	Output
Steam enthalpy (kJ)	1.3446E+06	2.1941E+06	2.1941E+06	2.6558E+06
	Output−input = 8.4945E+05		Output−input = 4.6170E+05	
	Total energy for two preheater units = 8.4945E+05 + 4.6170E+05			
	= 1.3112E06 kJ			

For the enthalpy calculations carried out on the syngas production, the majority of them have been carried out using the appropriate *component polynomial expressions* given in Appendix 3. If these are consistent with the assumptions made in Appendix 3, they can be employed here and should give values comparable to those calculated from the C_p values. Any differences should be down to the data sources used.

The relevant enthalpy polynomials were obtained from Appendix 3 and the input and output temperatures expressed in K.

The data values found are summarised based on the compositions stated for the natural gas stream where the enthalpies for this stream are calculated at the inlet and

TABLE 4.2

Enthalpy Calculations for Preheater E101B (see Summary Table in previous page)

Component	Initial Moles (kmol)	Input Component Enthalpy (kJ/kmol)	Input Stream Enthalpy (kJ)	Output Component Enthalpy (kJ/kmol)	Output Stream Enthalpy (kJ/kmol)
CH_4	94.68	2.1481E+04	2.0338E+06	2.5962E+04	2.4581E+06
C_2H_6	3.00	3.068E+04	9.2040E+04	3.8341E+04	1.1502E+05
C_3H_8	0.50	4.2352E+04	2.1176E+04	5.3463E+04	2.6732E+04
C_4H_{10}	0.40	5.4174E+04	2.1670E+04	6.7634E+04	2.7054E+04
N_2	1.40	1.7829E+04	2.4961E+04	2.0321E+04	2.8449E+04
H_2S	0.02	2.083E+04	4.166E+02	2.4093E+04	4.819E+02
Total			2.194085E+06		2.6558E+06

outlet temperatures for the stream indicated in the block diagram which have been stated previously as follows:

Item	E101A		E101B	
	Input	Output	Input	Output
Temperature (°C)	155	328	328	410

The input temperature is fixed at 155°C because a 5°C temperature drop has been allowed on the stream exiting the compressor.

4.6 ENERGY VALUES CALCULATED USING ENTHALPY POLYNOMIALS

The values calculated using the specific heat capacity polynomials and the enthalpy polynomials agree within 0.1%.

TABLE 4.3

Summary of Enthalpy Calculations for Preheaters E101A and E101B Using Enthalpy Polynomials

Item	E101A		E101B	
	Input	Output	Input	Output
Stream enthalpy (kJ)	−5.1605E+06	−4.3152E+06	−4.3152E+06	−3.8506e+06
	Output–input = 8.4530E+05		Output–input = 4.6460E+05	
	Total energy for two preheater units = 8.4530E+05 + 4.6460E+05			
	= 1.3099E+06 kJ			

The preheater units receive energy from a flue gas stream. The composition of the flue gas stream has been found as follows:

Component	kmol
CO_2	56.71
H_2O	110.57
N_2	485.28
O_2	16.83

4.7 FLUE GAS ENERGY CALCULATIONS

For this stream, the enthalpy polynomials can be used and the stream enthalpies determined based on the components and masses.

As part of the original flue gas calculations, the flue gas temperatures were fixed as follows:

	E101A		E101B	
Item	Input	Output	Input	Output
Temperature (°C)	341	305	662	644

The much higher temperatures in the item E101B are due to the set-up of the flue gas stream. Between the flue gas entering E101B and E101A, a further heat transfer operation occurs. This will be part of Chapter 6.

Using the temperature data set out for the flue gas, enthalpy calculations can be carried out.

4.7.1 FOR FEED PREHEATER UNIT E101A ON THE BLOCK DIAGRAM

The enthalpies in the following two tables have been calculated from the enthalpy polynomials. The component enthalpy figures have been truncated, but the totals are presented with the full number of significant figures.

We can write:

$$\text{Output} - \text{Input} = -8.3100\text{E} + 05\,\text{kJ}$$

In the same way, we can produce the data for *feed preheater unit E101B* on the block diagram.

We can write:

$$\text{Output} - \text{Input} = -4.5500\text{E} + 05$$

The total output from the flue gas is $-8.3100\text{E}+05 - 4.5500\text{E}+05 = -1.2860\text{E}+06\,\text{kJ}$.

TABLE 4.4

Input and Output Flue Gas Masses and Enthalpies for First Feed Preheater Unit

Component	kmol	Input		Output	
		kJ/kmol	Stream Enthalpy	kJ/kmol	Stream Enthalpy
CO_2	56.71	−3.6989E+05	−2.0977E+07	−3.7158E+05	−2.1072E+07
H_2O	110.57	−2.1818E+05	−2.4124E+07	−2.1948E+05	−2.4268E+07
N_2	485.28	1.8787E+04	9.1167E+06	1.7607E+04	8.5445E+06
O_2	16.83	1.8385E+04	3.0942E+05	1.7229E+04	2.8996E+05
Sum			−3.5675E+07		−3.6506E+07

TABLE 4.5

Input and Output Flue Gas Masses and Enthalpies for Second Feed Preheater Unit

Component	kmol	Input		Output	
		kJ/kmol	Stream Enthalpy	kJ/kmol	Stream Enthalpy
CO_2	56.71	−3.5322E+05	−2.0031E+07	−3.5421E+05	−2.0087E+07
H_2O	110.57	−2.0583E+05	−2.2759E+07	−2.0656E+05	−2.2839E+07
N_2	485.28	2.9720E+04	1.4423E+07	2.9085E+04	1.4114E+07
O_2	16.83	2.9080E+04	4.8942E+05	2.8463E+04	4.7904E+05
Sum			−2.7878E+07		−2.8333E+07

If the values of the required energy from the heat capacity and polynomial calculations are considered, the values obtained were:

Heat capacity enthalpy calculation for natural gas stream = 1.3112E06 kJ
Enthalpy polynomial calculation for natural gas stream = 1.3099E+06 1 kJ

With the input and output temperatures fixed for the flue gas stream, the calculated energy transferred from the stream was calculated as −1.2860E+06 kJ. This indicates that there is an ~2% difference in the energies available and transferred. For the preheater, the energy balance could be used to fix an output temperature from the preheater or more likely the exit temperature of the flue gas. The flue gas output temperature has been fixed by other operations, so the balance demonstrated here is considered acceptable. Some of the energy balancing on the flue gas stream is considered in Chapter 6.

SUMMARY OF IMPORTANT PARAMETERS
CALCULATED FOR PREHEATER

Component enthalpies were calculated using C_P data as a function of temperature.

The enthalpy values using C_P data and the enthalpy polynomial calculations were used.

With the input and output temperatures fixed for the flue gas stream, the calculated energy transferred from the stream was calculated as $-1.2860E+06$ kJ using enthalpy polynomials.

The total energy requirement for the two preheater units $= 8.4530E+05 + 4.6460E+05 = 1.3099E+06$ kJ based on the enthalpy polynomials.

The total energy requirement for the two preheater units $= 8.4945E+05 + 4.6170E+05 = 1.3112E06$ 3112E06 kJ based on C_P polynomials.

4.8 DESULPHURISATION OF NATURAL GAS STREAM (ITEM V101 ON THE BLOCK DIAGRAM)

REQUIRED

Equipment for removal of sulphur from the natural gas stream by adsorption.
Calculation of amount of adsorbent (zinc oxide) to remove sulphur.

The natural gas is now almost ready for processing in the primary reformer. Before entering the reformer, the stream has to have the small amount of sulphur-bearing hydrogen sulphide removed. If this was present on entering the primary reformer, the reformer catalyst would be poisoned.

There are alternatives for the removal of the hydrogen sulphide, but this text recognises that the amount of hydrogen sulphide is relatively small and it will be removed by a method that has been employed in the production of the syngas for a number of years, adsorption using zinc oxide.

The process is treated simply here. The only sulphur present is in the hydrogen sulphide which is present in small amounts. This can be handled by employing a zinc oxide bed through which the natural gas passes and the hydrogen sulphide is adsorbed.

For our purposes, based on prior experience, we fix the zinc oxide bed as having a 19.8% pickup of the sulphide. To work out the amount of the oxide required, we require the mass of the sulphur to be adsorbed.

Per 100 kmol of the natural gas stream, we have 0.02 kmol of the hydrogen sulphide. Thus in the molecule H_2S, with a relative atomic mass of sulphur 32, the amount of sulphur to be removed is:

$$0.02 \times 32.0 = 0.64 \, \text{kg}.$$

Normally, in the adsorbing system, the stream would be flowing. For the purposes of illustration, we assume a flow of kmol/s. The zinc oxide bed would have a definite active life, which we will assume to be 1 year. On that basis, we consider an active time period in seconds:

$$365 \times 24 \times 60 \times 60 = 31,536,000 \text{ s}$$

Realistically, the operating time reduces the factor $365 \times 24 = 8760$–8000 to allow for operating downtimes. That then gives an active period of:

$$8000 \times 3600 = 28,800,000 \text{ s}$$

For this stated period of operation, the mass of zinc oxide required would be:

$$\frac{m}{100} = 28,800,000 \times \frac{0.641}{0.198}$$
$$m = 9.32 \times 10^7 \text{ kg/s}$$

With such a large feedstock flow rate, the zinc oxide requirement would almost certainly have to be met by using more than one bed.

SUMMARY OF IMPORTANT PARAMETERS
CALCULATED FOR DESULPHURISATION

On a per-second basis, a zinc oxide bed removing 19.8% sulphur for the mass (m) of zinc oxide supplied is considered. On the basis of 100 kmol feed, the amount of zinc is calculated as m, where:

$$m = 9.32 \times 10^7 \text{ kg/s}$$

If such a quantity of feed was to be processed on this time basis, multiple zinc oxide beds would be employed.

5 Problems to Solve with Hints

5.1 QUESTIONS RELATED TO SYNGAS AND AMMONIA GAS PRODUCTION

You will find more 'orthodox' questions in this chapter but questions based on the detailed analysis you should have carried out as part of the mass and energy balances for the ammonia syngas.

There is still a need for basic understanding of the thermodynamics and principles of balancing.

Problem 1:
In the production of the ammonia syngas described in the text, there is an imperative to keep the nitrogen-to-hydrogen ratio as close as possible to the stoichiometric ratio, i.e. 1:3. Some students (and non-students) do seem to have problems handling a non-stoichiometric feed and producing the composition of an output stream. The main thrust of this book has been to produce the mass and energy balances for ammonia syngas production.

As a useful exercise, it helps to consider what happens to the syngas in the ammonia synthesis loop. The stoichiometric ammonia synthesis equation is considered to be:

$$N_2 + 3H_2 \leftrightarrow 2NH_3 \quad \Delta H - ve$$

Even using a basic approach to thermodynamics as enshrined in Le Chatelier's principle, it will be recognised that in equilibrium terms, the pressure should be as high as feasible and, given the exothermic nature of the reaction, the temperature should be as low as possible. This is a dilemma often faced with exothermic reactions. In kinetic terms, we usually find that the higher the temperature, the faster the reaction. In terms of the final equilibrium achieved, we have stated that we want as low a temperature as possible. There is a compromise to be reached in terms of the temperature fixed. This is also a case where an efficient catalyst would be used. Such catalysts exist but are sensitive to the presence of carbon oxides. You will have observed that in mass balancing on the syngas, there is considerable effort made to remove carbon oxides.

In producing ammonia for this question, we are operating at 200 bar (many modern plants can operate at this pressure, but the practical pressure is often set lower). The temperature has been set at 450°C. The challenge is simply that the synthesis gas has a composition that is equimolar in nitrogen and hydrogen. What is the output composition from the reactor assuming equilibrium is reached?

The data available on K_p for this reaction indicate that at the operating temperature, the value of K_p is 1.44×10^{-4}.

Problem 2:
Hydrogen can be produced by reacting methane. For the purposes of this calculation, the pressure is set at 1 bar and the reaction is considered to be:

$$CH_4 + H_2O \leftrightarrow CO + 3H_2$$

For the reactants, their initial ratio H_2O/CH_4 is fixed at 6.0; i.e., as discussed for the primary reformer in the text, it is useful to increase the amount of steam as a diluent and to supress unwanted side reactions. For the purposes of this calculation, the reaction is allowed to go to equilibrium. It is found that a fractional conversion of methane of 0.5 is achieved; calculate the temperature at which the reaction was carried out.

All the necessary K_p values are contained in Appendix 1.

Problem 3:
As part of ammonia production, it is necessary to produce a synthesis gas containing stoichiometric proportions of nitrogen and hydrogen. As you will have observed, one of the first stages in the synthesis gas production is carried out in a reformer where methane gas can be considered converted to hydrogen according to the equations:

$$CH_4 + H_2O \leftrightarrow CO + 3H_2 \qquad (5.1)$$

$$CO + H_2O \leftrightarrow CO_2 + H_2 \qquad (5.2)$$

These were the equations used in the text for the primary reformer, and an analysis is given in the text for the reactions in terms of fractional conversions z_{CH4} and z_{CO}.

For the purposes of this calculation, the reformer is operated at a pressure of 1 bar (obviously lower than in practice) and a temperature of 900 K. The methane and steam are the only initial reactants and are fed in the ratio $H_2O/CH_4 = 3.5{:}1$.

You are asked to confirm that under these conditions, the fractional conversion of methane in the first reaction should be 89.2% and for the second reaction should be 59.6%. The K_p values for these reactions are available in the appropriate appendix, and the values should be found at the quoted temperature. A solution can be obtained using an appropriate convergence procedure (Excel SOLVER was used in this case).

Problem 4:
In teaching the energy balance, one of the classic problems set is trying to find a reactor exit stream temperature when the reactor operates adiabatically. In this type of problem, it is often useful to use mean molar heat capacities. These can be calculated but are often tabulated and can be referenced. If these heat capacity figures are

available, then calculations can be carried out and a final temperature reached using the simplest energy balance for the reactor:

$$\text{Input Energy} = \text{Output Energy}$$

This requires the output temperature to be known, but that is the parameter being calculated; hence, a number of iterations are required. For each iterated temperature, an appropriate value of the mean molar heat capacity is required.

In Appendix 3, various calculation methods for calculating enthalpy changes are outlined. The most convenient form of enthalpy data available in the appendix for this calculation is to use the polynomial expressions where a base temperature of 0 K is used. For each component, the appropriate polynomial expression is available. To carry out the balance, the input temperature can be taken as 25°C (298 K). It is the output temperature that has to be calculated, and this can be done using the polynomial expressions.

It should be understood that to do the energy balance, the reactor mass balance should also be known. In the problem, the input fuel composition is known as is the composition of the exit gas after the fuel has been burned. For such a problem, the stoichiometric equations will not be fully known, and using a mass balance fundamental, input atom species = output atom species, it will be possible to carry out the mass balance over the reactor. Once the mass balance is established, it is proposed that the energy balance can be written in terms of enthalpies using the appropriate species polynomials:

$$\sum_{1}^{N} m_i \left(a + bT_i + cT_i^2 + dT_i^3 \right) = \sum_{1}^{N} m_o \left(a + bT_o \right) + cT_o^2 + dT_o^3$$

where subscript i refers to an input species; subscript o refers to an output species; m is a species mass; T is a temperature in K; a, b, c and d are enthalpy polynomial coefficients; and N is the number of component species.

Having produced the energy balance, the balance needs to be solved for the temperature. This can be done using, for example, MATLAB or Excel.

Problem
One mole of a fuel gas used to raise steam has the following composition:

Component	Mole Fraction
CO	0.21
CO_2	0.071
H_2	0.286
CH_4	0.074
N_2	0.357

This gas was mixed with air and reacted adiabatically. After combustion, the exhaust stream had a composition measured as:

Component	Mole Fraction
CO	0.043
CO_2	0.0.15
H_2	0.0.086
CH_4	0.0.021
N_2	0.0.699

Work out a ratio of fuel gas to air.

If the fuel gas and air entered the burner at 20°C, what is the adiabatic flame temperature?

If you are not familiar with this term, then it is important to find out and return to this problem when you are more knowledgeable about the term. Dependent on how you intend to solve the problem, you will find the polynomials for enthalpies of the various species given in Appendix 3 to be useful.

The problems you have so far dealt with in this appendix are specifically linked with aspects of the problems you have dealt with in the text dealing with the balances related to the preparation of ammonia syngas.

It is obvious from the production of the syngas that you have to be able to handle the challenge of:

$$Input = Output$$

at equilibrium or steady state.

The challenge posed by the need to consider reactions approaching or at equilibrium requires the equilibrium parameter related to the second law of thermodynamics, the equilibrium constant, to be used and the necessary simple mathematics for solution to be available.

There are certain aspects of the use of the equilibrium constant, and its form for a particular reaction, that allows decisions to be made about possible processing conditions for the reaction.

The following problems illustrate the use of the relatively simple equation for the formation of ammonia from a simple syngas. The problems simply ask for application of basic analysis of the reaction, what the implications of the analysis are and how then to fix conditions as dictated by the balance and thermodynamic calculations.

Problem 5:

    a. If a 1:3 mixture of nitrogen and hydrogen is fed to an ammonia synthesis reactor operated so that 20% of the nitrogen is converted to ammonia, what will be the composition of gas leaving the reactor?

    b. Suppose that the reactor operates at a total pressure of 150 bar. What will be the partial pressures of each of the components in the exit stream? Indicate the value of K_P. Try to find the temperature of operation at which the conversion will be achieved.

The following expression can be used for ammonia synthesis:

$$\ln K_P = \frac{12623.84}{T} - 27.471$$

T is in K.

Problem 6:

Suppose an ammonia synthesis reactor is operated at 200 bar and 454°C. What is the fractional conversion of a stoichiometric mixture of nitrogen and hydrogen? Also, calculate the fractional conversion at 500 bar and 454°C and 200 bar and 399°C.

a. What conclusions could be drawn about the effect of temperature and pressure on the fractional conversion?
b. Are these conclusions what you expected?
c. When the ammonia production reactor is operated, it is usually necessary to operate a recycle stream for the return of unreacted nitrogen and hydrogen. Suppose, operationally, a build-up of inerts in the recycle stream has been allowed instead of being purged; the inerts have reached a point where the level of nitrogen:hydrogen:inerts is 1:3:0.5 kmol. It is unlikely that in normal operations, inert material would rise to this level, but it is a useful exercise to investigate the effect of the inert diluent. If the reactor operates at 200 bar and 454 K, calculate the effect of the presence of the inerts and comment on the result.

5.2   PROBLEMS RELATING TO SYNGAS PRODUCTION NOT INCLUDED IN THE MAIN TEXT

This appendix contains a certain number of problems in mass and energy balancing related to the ancillary equipment associated with the production of ammonia syngas. In carrying out the synthesis of the gas, there are a number of items of equipment that require energy inputs. An examination of the block diagram and the relevant calculations indicates that the primary reformer (Item F101) is a major item requiring a net energy input. The large demand for energy as heat imposed by the endothermic methane–steam reaction is usually met by putting banks of tubes, packed with the necessary catalyst into, what is essentially, a furnace, with the process gas flowing through the tubes.

The energy required for the furnace is supplied by burning a fuel in air that uses the energy from the exothermic combustion reaction to heat the furnace. Obviously, in a text such as this, no attempt is made to look in detail at the furnace and associated equipment design, but the fuel used is the same natural gas used as the syngas feedstock. The burning of the fuel for the primary reformer energy produces a flue gas that has a high temperature and can potentially be used to supply energy to other equipment. This has already been illustrated in the mass and energy balances for the preheater used in the syngas production. There are some other items before the flue gas is sent to a stack that can be considered.

The flue gas stream can be used to transfer energy to other items of equipment, not necessarily part of the syngas production. In this particular set of problems, there is a challenge to use the flue gas energy to produce superheated steam. There is also a mass and energy balance associated with the superheater where there has to be a radiant shield. The radiant shield associated with a boiler protects the following superheater from transient conditions where there could be damage to the super-heater metal.

Problem 1:

It is known that energy from the flue gas stream will be used to produce steam from an existing steam/water system that exists under the following conditions:

Steam/Water	Input	Output
Temperature (°C)	278	278
Pressure	62	62

At the point where the flue gas stream enters the radiant shield heater, it has the following properties:

Flue Gas	Input	Output
Temperature (°C)	950	892

To calculate the energy available from the flue gas, it is necessary to do an energy balance. Enthalpies can be calculated using the enthalpy polynomials available in Appendix 3.

The available energy should be calculated as:

$$-1,563,257.67 \, kJ$$

Realistically, you would expect to lose about 2% of this energy, so the energy available for steam raising will be:

$$1.532 \times 10^6 \, kJ.$$

Using steam tables to find the latent heat at vaporisation at 62 bar and 278°C, we find a value of

$$1557 \, kJ \, / \, kg$$

You should find that the amount of steam raised is

$$984\,\text{kg}.$$

A detailed solution to this problem is given as Solution 1 in Chapter 6.

Problem 2:

The flue gas from the radiant stream boiler now enters a superheater, and more energy is exchanged from the flue gas to raise steam. The amount of steam entering the superheater is 9704 kg at the conditions shown below:

Steam/Water	Input	Output
Temperature (°C)	278	420
Pressure	62	58.5

The pressure drop through the superheater is incorporated. This will depend on the actual design layout of the heat exchanger system supplying the energy.

At the point where the flue gas stream enters the superheater, it has the following entry and exit conditions:

Flue Gas	Input	Output
Temperature (°C)	892	?

The mass balance will not alter.

This situation will require an energy balance to be carried out. Often in these situations, the energy balance is used to fix an output temperature. In this case, you are asked to carry out the energy balance and fix the flue gas exit temperature.

You should find a temperature of 726.9°C.

6 Answers to Problems in Chapter 5

6.1 ANSWERS TO QUESTIONS RELATED TO SYNGAS AND AMMONIA GAS PRODUCTION

Q1 Solution:

We are given the stoichiometric equation:

$$N_2 + 3H_2 \leftrightarrow 2NH_3$$

The text uses a fractional conversion, and this is consistently used to work out the reaction mass balance. As this approach is usually adopted, draw up an equilibrium balance table:

Basis: 1 kmol

Component	Initial (kmol)	Equilibrium (kmol)
N_2	1.0	$1 - z_{N_2}$
H_2	1.0	$1 - 3z_{N_2}$
NH_3	-	$2z_{N_2}$
Total	2.0	$2 - 2z_{N_2}$

$$z_{N_2} = \frac{\left(\text{Initial moles } N_2 - \text{Equilibrium moles } N_2 \right)}{\left(\text{Initial moles } N_2 \right)}$$

Thus, the equilibrium moles of N_2 can be written as:

$$1 - 1 \times z_{N_2} = 1 - z_{N_2}$$

Obviously, (initial moles − equilibrium moles) represents the amount of N_2 converted, and this is equal to z_{N_2}

If z_{N_2} moles of N_2 are converted, then $3z_{N_2}$ moles of H_2 are converted.

Thus, at equilibrium, there will be $1 - 3z_{N_2}$ moles of H_2.

On the same basis, if z_{N_2} moles of N_2 are converted, then $2z_{N_2}$ moles of ammonia must be produced.

If we write K_p as:

$$K_p = \frac{p_{NH_3}^2}{p_{N_2}\, p_{H_2}^3}$$

as explained in the main text, we can write:

$$p_i = y_i P$$

where p_i is the partial pressure of component 'i', y_i is the gas/vapour mole fraction of 'i' and P is the total pressure.

Thus, using the data from equilibrium mass balance and dropping the subscripts:

$$K_p = 1.44 \times 10^{-4} = \frac{(2z)^2 (2-2z)^2}{(1-z)(1-3z)^3 \times 200^2}$$

You will note that the expression $(2 - 2z)^2$ appears because each of the component moles has to be divided by the total equilibrium moles. This equation can be solved graphically, but it is better to use systems such as MATLAB or Excel.

Using Excel, the equation was written as:

$$1.44 \times 10^{-4} - \frac{(2z)^2 (2-2z)^2}{(1-z)(1-3z)^3 \times 200^2} = 0$$

Using SOLVER, the value of z can be calculated as 0.189.

This gives the final table:

Component	Initial (kmol)	Equilibrium (mol)
N_2	1.0	$1 - 0.189 = 0.811$
H_2	1.0	$1 - 3 \times 0.189 = 0.433$
NH_3	-	$2 \times 0.189 = 0.378$

Obviously, H_2 being a limiting component, we can check what happens if H_2 is present in stoichiometry proportions. We find that z_{N_2} becomes 0.57, and hence, the amount of ammonia produced is 1.14 kmol.

Q2 Solution:

The reaction involved is the methane–steam reaction:

$$CH_4 + H_2O \leftrightarrow CO + 3H_2$$

If we draw up a simple equilibrium balance taking 1.0 kmol of methane as basis, then we can say:

Component	Initial (kmol)	Equilibrium (kmol)
CH_4	1.0	$1 - z_{CH4}$
H_2O	6.0	$6 - z_{CH4}$
CO	-	z_{CH4}
H_2	-	$3z_{CH4}$
Total	7	$7 + 2z$

Taking $z_{CH4} = \dfrac{(\text{Initial moles } CH_4 - \text{Equilibrium moles } CH_4)}{(\text{Initial moles } CH_4)}$

$$z_{CH4} = \frac{(1-x)}{(1)}$$

$$x = 1 - z_{CH4}$$

At the same time, we recognise that amount of CH_4 converted must be $1 - x = z_{CH4}$

Using this and the reaction stoichiometry, the moles H_2O converted must be z_{CH4} and the moles of H_2O at equilibrium must be $6 - z_{CH4}$.

Again taking the reaction stoichiometry, the amount of CO produced must be z_{CH4} and the amount of H_2 produced must be $3z_{CH4}$.

Taking the sum of the equilibrium moles, we find an expression: $7 + 2z_{CH4}$.

If we express K_p in terms of component partial pressure 'p_i' and write $p_i = y_i\,P$, where p_i is the partial pressure, y_i the component mole fraction and P the total pressure:

$$K_p = \frac{p_{CO}\; p_{H2}^3}{p_{CH4}\; p_{H2O}} = \frac{y_{CO}\; P\left(y_{H2}\, P\right)^3}{y_{CH4}\; P\; y_{H2O}P} = \frac{y_{CO}\; y_{H2}^3\; P^2}{y_{CH4}\; y_{H2O}}$$

Writing each mole fraction as component moles/total moles:

$$\frac{\left(\dfrac{z_{CH4}}{7+2z_{CH4}}\right)\left(\dfrac{3z_{CH4}}{7+2z_{CH4}}\right)^3}{\left(\dfrac{1-z_{CH4}}{7+2z_{CH4}}\right)\left(\dfrac{6-z_{CH4}}{7+2z_{CH4}}\right)} = \frac{z_{CH4}\left(3z_{CH4}\right)^3}{\left(1-z_{CH4}\right)\left(6-z_{CH4}\right)\left(7+2z_{CH4}\right)^2}$$

For a value of $z_{CH4} = 0.5$, we find a value of $K_p = 9.59 \times 10^{-3}$.

We know $\ln K_p = \ln(9.59 \times 10^{-3}) = -4.641$.

$$\ln K_p = \frac{-27,106}{T} + 30.42$$

$$T = \frac{-27,106}{(-4.641 - 30.42)} = \frac{-27,106}{-35.061} = 773.1K$$

Q3 Solution:

The important parts of the book text to access are the K values and the equations in Appendix 3.

The other important table to access is the table that gives the primary reformer analysis in terms of two fractional conversions z_{CH_4} for reaction and z_{CO} for Reaction 2. There is an explanation of the primary reformer analysis given in Appendix 2.

The analysis carried out, in effect, assumes that the thermodynamics is dependent on the initial and final states only and not the path taken.

The analysis finds the equilibrium products for Reaction 1 in terms of z_{CH_4} and taking the products from this reaction as the starting amounts for Reaction 2, does an analysis in terms of z_{CO}.

Taking the book text table that analyses the primary reformer, we can follow its logic and adopt the figures appropriate to this problem.

Assume equal moles of CH_4 and H_2O.

Initial moles of $CH_4 = 1$ kmol and $H_2O = 3.5$ kmol.

On this basis, we can write the equilibrium mass balance table:

Component	Initial Moles	Moles after Reaction 1	Moles after Reaction 2
CH_4	1.0	$1 - z_{CH_4}$	$1 - z_{CH_4}$
CO	-	z_{CH_4}	$z_{CH_4} - z_{CH_4} z_{CO}$
H_2O	1.0	$3.5 - z_{CH_4}$	$(3.5 - z_{CH_4}) - z_{CH_4} z_{CO}$
CO_2	-	-	$z_{CH_4} z_{CO}$
H_2	-	$3 z_{CH_4}$	$3 z_{CH_4} + z_{CH_4} z_{CO}$
Total	2.0		

At the temperature stated, K_{p1} for Reaction 1 can be calculated as 1.352 and K_{p2} for Reaction 2 as 2.280, thus writing:

$$K_{P1} = \frac{p_{CO}\, p_{H_2}^3}{p_{CH_4}\, p_{H_2O}} = \frac{y_{CO}\left(y_{H_2}\, P\right)^3}{y_{CH_4}\, y_{H_2O}}$$

$$K_{P2} = \frac{y_{CO_2}\, y_{H_2}}{y_{CO}\, y_{H_2O}}$$

using the fact that $P = 1$.

The appropriate values of y can be written in terms of z_{CH_4} and z_{CO}:

$$K_{P1} = \frac{\left(z_{CH_4} - z_{CH_4} z_{CO}\right)\left(3 z_{CH_4} + z_{CH_4} z_{CO}\right)^3}{\left(1 - z_{CH_4}\right)\left(3.5 - z_{CH_4} - z_{CH_4} z_{CO}\right)\left(4.5 + 2 z_{CH_4}\right)^2} \qquad (6.1)$$

$$K_{P2} = \frac{\left(z_{CH_4} z_{CO}\right)\left(3 z_{CH_4} + z_{CH_4} z_{CO}\right)}{\left(z_{CH_4} - z_{CH_4} z_{CO}\right)\left(3.5 - z_{CH_4} - z_{CH_4} z_{CO}\right)} \qquad (6.2)$$

Essentially to answer the question, we need to solve for z_{CH_4} and z_{CO}, two equations and two unknowns in this case.

This was done using SOLVER in Excel, to find a function to set to zero.

The equation for K_{P1} was multiplied by K_{P2}/K_{P1}. The resulting equation was then subtracted from equation (6.2) and the values of z_{CH_4} and z_{CO} calculated using SOLVER to find the z values that made this function zero.

SOLVER gave a solution, and it should be noted that the solution is sensitive to the initial guessed values of z_{CH_4} and z_{CO}. It is also important to constrain $0 < z_{CH_4} < 1$ and $0 < z_{CO} < 1$.

Values of 0.892 and 0.596 are obtained as specified in the question.

Q4 Solution:

Take a basis of 100 kmol fuel gas.
We are given input and output compositions.
A mass balance must be carried out; no reaction equations are specified.
This is a classic situation where atom balances must be carried out.
An input/output table can be drawn up:

| | Input Fuel Gas | | | | Output Gas (Dry) W | | |
| | | Amount | | | | Amount | |
Component	Composition	(kmol)	C-atoms	Air	Composition	(kmol)	C-atoms
CO	21.0	21.0	21.0		4.3	0.043 W	0.043 W
CO$_2$	7.1	7.1	7.1		15.0	0.15 W	0.15 W
H$_2$	28.6	28.6			8.0	0.08 W	
CH$_4$	7.4	7.4	7.4		2.1	0.021 W	0.021 W
N$_2$	35.7	35.7		79.0	63.9	0.639 W	
O$_2$				21.0			
H$_2$O							
Total		100	35.5				0.214 W

The following balances can be carried out:
Carbon atom balance:

Carbon atoms in = Carbon atoms out
$35.5 = 0.214\ W$

This can be solved to give W, the kmol of the output stream on a dry basis.
Nitrogen atom balance:

Nitrogen in fuel gas + Nitrogen in air = Nitrogen in output gas
$35.7 + \text{Nitrogen in air} = 0.699 \times 165.89$
Nitrogen in air = 80.26 kmol
Thus, O$_2$ in air $= 80.26 \times (21/79) = 21.33$ kmol

Hydrogen atom balance:

H_2 in fuel gas = H_2 in output gas + H_2O in output gas
$28.6 \times 2 + 7.4 \times 4 = 0.086 \times 165.89 \times 2 + 0.021 \times 165.89 \times 4 + H_2O \times 2$ in output gas

Thus, H_2 in output gas:
$86.8 = 28.53 + 13.94 + 2H_2O$ in output gas
H_2O in output gas = 22.17 kmol.
It is now possible to write a summary:

With 165.89 kmol of output gas
$CO = 165.89 \times 0.043 = 7.13$ kmol
$CO_2 = 165.89 \times 0.15 = 24.88$ kmol
$H_2 = 165.89 \times 0.086 = 14.27$ kmol
$CH_4 = 165.89 \times 0.021 = 3.48$ kmol
$N_2 = 165.89 \times 0.699 = 115.96$ kmol
Total = 165.72 kmol

In the input air, the N_2 has been calculated as 80.26 kmol and O_2 as 21.33 kmol.

Thus, the ratio of fuel gas to air must be $\dfrac{100}{(80.26 + 21.33)} = 0.984$.

    b. As indicated in the question, finding the flame temperature requires a statement of solution of the adiabatic energy balance.

For the input, all the values of m_i have already been calculated.
    The temperature T_i is specified (298 K), so the input energy can be calculated using the polynomial equations.
    This should give a total energy input of 4.201×10^6 kJ.
    In the output stream, the output temperature is the unknown.
    All the values of m_o have been found in the equilibrium mass balance.
    The individual enthalpies can be found from the polynomials.
    SOLVER was used to set a function
    Input energy − Output energy to zero by altering the value of T_o
    The value of $T_o = 1725.9$ K was found.

Q5 Solution:
    a. Basis: 1 kmol

$$\text{Reaction: } N_2 + 3H_2 \leftrightarrow 2NH_3$$

$$z_{N_2} = \frac{(1.0 - \text{Equilibrium moles } N_2)}{(1.0)}$$

Therefore, equilibrium moles $N_2 = 1 - z_{N_2}$

Draw up a mass balance table at equilibrium:

Component	Initial (kmol)	Equilibrium (kmol)
N_2	1.0	$1 - z_{N_2}$
H_2	3.0	$3 - 3z_{N_2}$
NH_3	-	$2z_{N_2}$
Total	4.0	$4 - 2z_{N_2}$

Take compositions in mole fractions:

$$\text{Mole fraction of } N_2 = \frac{1 - z_{N_2}}{4 - 2z_{N_2}}$$

$$\text{Mole fraction of } H_2 = \frac{3 - 3z_{N_2}}{4 - 2z_{N_2}}$$

$$\text{Mole fraction of } NH_3 = \frac{2z_{N_2}}{4 - 2z_{N_2}}$$

The question states $z_{N_2} = 0.2$

Thus:

$$\text{Mole fraction of } N_2 = \frac{1 - z_{N_2}}{4 - 2z_{N_2}} = \frac{0.8}{3.6} = 0.222$$

$$\text{Mole fraction of } H_2 = \frac{3 - 3z_{N_2}}{4 - 2z_{N_2}} = \frac{2.4}{3.6} = 0.667$$

$$\text{Mole fraction of } NH_3 = \frac{2z_{N_2}}{4 - 2z_{N_2}} = \frac{0.4}{3.6} = 0.111$$

b. Total pressure = 150 bar

Take $y_i P = p_i$, where p_i is the partial pressure of component 'i', P is the total pressure and y_i is the mole fraction of component 'i'.
We know $P = 150$ bar.
Thus:

$$p_{N_2} = 0.222 \times 150 = 33.3 \, \text{bar}$$

$$p_{H_2} = 0.667 \times 150 = 100.05 \, \text{bar}$$

$$p_{NH_3} = 0.111 \times 150 = 16.65 \, \text{bar}$$

We can write:

$$K_p = \frac{p_{NH_3}^2}{p_{N_2}\, p_{H_2}^3} = \frac{16.65^2}{33.3 \times 100.05^3} = 8.313 \times 10^{-6}$$

To find a temperature, use the given equation:

$$\ln K_p = \frac{12{,}623.84}{T} - 27.471$$

$$T = \frac{12{,}623.84}{\ln K_p + 27.471}$$

Substituting in K_p, the value of temperature can be found to be 800.3 K = 527.3°C.

Q6 Solution:

$$\text{Reaction: } N_2 + 3H_2 \leftrightarrow 2NH_3$$

For $T = 454°C$ (727 K) and $P = 200$ bar:

Component	Initial (kmol)	Equilibrium (kmol)
N_2	1.0	$1 - z_{N_2}$
H_2	3.0	$3 - 3z_{N_2}$
NH_3	-	$2z_{N_2}$
Total	4.0	$4 - 2z_{N_2}$

At $T = 454°C$ (727 K), use the equation for K_p:

$$\ln K_p = \frac{12{,}623.84}{727} - 27.471$$

$$K_p = e^{-10.107} = 4.08 \times 10^{-5}$$

Thus, at 727 K:

$$4.08 \times 10^{-5} = \frac{\left(\dfrac{2z_{N_2}}{4 - 2z_{N_2}}\right)^2 \cdot 200^2}{\left(\dfrac{1 - 2z_{N_2}}{4 - 2z_{N_2}}\right) \cdot 200 \cdot 200^3}$$

$$4.08 \times 10^{-5} = \frac{(2z)^2 \cdot (4 - 2z)^2}{(1 - z)(3 - 3z)^3 \cdot 200^2}$$

We can write this as:

$$4.08 \times 10^{-5} - \frac{(2z)^2 \cdot (4-2z)^2}{(1-z)(3-3z)^3 \cdot 200^2} = 0$$

We can use SOLVER or Goal Seek in Excel.

Find $z = 0.387$

For the same stoichiometry and temp

$P = 500$ bar

Find $K = 4.08 \times 10^{-5}$ using Goal Seek

$z = 0.56$

For the same stoichiometry but temperature $= 399°C$ (672 K) with $P = 200$ bar

Find $z = 0.522$

a. Summarising:

	T	P	z
1	727	200	0.387
2	727	500	0.560
3	672	200	0.522

For Cases 1 and 2 with temperature constant
Pressure increases, and conversion increases
Increasing P increases z
    Cases 1 and 3 with pressure constant
As temperature reduces, conversion increases

b. These are consistent with Le Chatelier's principle. For exothermic reactions, any condition allows energy transfer from system to surroundings and pushes equilibrium to the right, and a reduced temperature achieves this.
    For a reaction where we go from four volumes on the left-hand side to two volumes on the right-hand side, a higher pressure pushes the equilibrium to the right. Cases 1–3 indicate this happens.
    In terms of the quantitative statement, it is easy to show that K_p increases as T reduces – z_{N_2} will increase as T reduces.

Inspection of equation for K_p in terms z_{N_2} indicates that if the pressure term is taken over to the other side of the equation, then z will increase as P increases.

c. For the conditions stated, $T = 454°C$ (727 K) and $P = 200$ atm. It has been calculated that $K = 4.08 \times 10^{-5}$. The equilibrium mass balance can be written:

Basis: 1 kmol

Component	Initial (kmol)	Equilibrium (kmol)
N_2	1.0	$1 - z_{N_2}$
H_2	3.0	$3 - 3z_{N_2}$
NH_3	-	$2z_{N_2}$
Inert	0.5	
Total	4.5	$4.5 - 2z_{N_2}$

The K_p can be written as:

$$K_p = \frac{\left(\dfrac{2z \times P}{4.5 - 2z} \right)^2}{\left(\dfrac{1-z}{4.5-2z} \times P \right) \left(\dfrac{3-3z}{4-2z} \times P \right)^3}$$

$$4.08 \times 10^{-5} = \frac{(2z)^2 (4.5 - 2z)^2}{(1-z)(3-3z)^3 P^2}$$

This can be solved using Goal Seek or an appropriate method. It can be shown that $z_{N_2} = 0.361$. There has been reduction in z. Numerically, the equilibrium mixture is now $(4.5 - 2z)$.

Mathematically, this reduces the value of z.

This dilution effect can sometimes used in reactions where the equilibrium reaction mixture volume has increased. Theoretically, an increased pressure would reduce z and a decrease in pressure is often not practical or desirable. This 'diluent' effect can sometimes be used in such cases.

6.2 ANSWERS TO PROBLEMS RELATING TO SYNGAS PRODUCTION NOT INCLUDED IN THE MAIN TEXT

Q1 Solution:

It is necessary to find the component enthalpies for the flue gas at the stated input and output conditions. The enthalpy polynomials are used and the following figures found:

Component	kmol	Input Stream Enthalpy kJ	Output Stream Enthalpy kJ
CO₂	56.71	−1.91E+07	−1.93E+07
Water	110.57	−2.14E+07	−2.17E+07
Nitrogen	485.28	1.96E+07	1.85E+07
Oxygen	16.83	6.60E+05	6.25E+05
		−2.03E+07	−2.19E+07
	Total	−20,322,487.73	−21,885,745.40

$$\text{Enthalpy Difference} = -1,563,257.67 \text{ kJ}$$

$$= -1.563 \times 10^6 \text{ kJ}$$

In the problem, it states that the enthalpy of vaporisation (latent heat) is:

$$1557 \text{ kJ/kg.}$$

Taking the energy being transferred from the flue gas to the process stream as the adjusted value for a realistic 2% energy loss, we obtain the figure:

$$1.532 \times 10^6 \text{ kJ}$$

Using the appropriate latent heat of the steam, we can calculate the kg of steam that can be generated:

$$= 1.532 \times 10^6 / 1557 = 984 \text{ kg}$$

Q2 Solution:
To fix the flue gas outlet temperature, the process energy balance is carried out.

Energy Balance:
Consider the steam stream fed into the superheater. The entering stream is saturated at a pressure of 62 bar. From steam tables, the corresponding saturation pressure can be read (interpolated) as 278°C. This steam passes through the superheater where energy from the flue gas transfers to the steam and exits in the outlet stream at a pressure of 58.5 bar. Steam tables indicate that the temperature corresponding to this at saturated stream conditions is 273.9°C. Thus,

$$420 - 273.9 = 146.1 \text{ degrees of superheat are present.}$$

From steam tables, the specific enthalpy of the saturated steam at inlet is read as 2782 kJ/kg.
The superheated steam at outlet at 420°C can be read as 3229 kJ/kg.

As stated, 9704 kg of steam enter the superheater; thus, the amount of energy required to raise the superheated steam at 420°C is:

$$(3229 - 2782) \times 9704 = 4.338 \times 10^6 \text{ kJ}$$

As has been demonstrated in previous calculations, we have to carry out an iterative process to fix the outlet temperature.

The inlet temperature for the flue gas is known to be 892°C. Using the enthalpy polynomials in Appendix 3, the component enthalpies can be calculated at 1165 K, and using the relevant component amounts, the stream enthalpies in kJ can be calculated as follows:

Composition		In
CO_2	56.71	−1.93E+07
Water	110.57	−2.17E+07
Nitrogen	485.28	1.85E+07
Oxygen	16.83	6.25E+05
		−2.19E+07

The energy required for the steam superheat stream is known. The energy balance can be written as:

$$\text{Steam Stream requirement} = 4.338 \times 10^6 \text{ kJ} = \left(\text{Energy Out} - 2.189 \times 10^7\right)$$

The energy out term for the flue gas can be written in terms of the enthalpy polynomials as:

$$\text{Stream Enthalpy} = \sum_{1}^{N} m_i H_i$$

where H_i is the component enthalpy represented by the relevant component enthalpies, m_i is the component moles and N is the number of components.

$$\text{In the form}: H_i = a + bT + cT^2 + dT^3$$

The relevant constants are found, as usual, in Appendix 3.

Each of the polynomial expressions carried terms in T, T^2 and T^3 where it is was then necessary to calculate the value of T.

As in a number of problems solved here, SOLVER in an Excel spreadsheet was set to produce an answer by setting a function of the form:

$$-4.338 \times 10^6 \text{ kJ} = \left(\text{Energy Out} - \left(-2.189 \times 10^7\right)\right)$$

$$-4.338 \times 10^6 - 2.189 \times 10^7 - \sum_{1}^{N} m_i H_i = 0$$

T, the temperature, is the unknown in the polynomial expressions. A value of 726.8°C is found.

Appendix 1
Basic Equations Relating to the Thermodynamics of Equilibrium Constants and Their Calculation and Their Values for the Methane–Steam Reaction and Water–Gas Shift Reaction

If you have come to this appendix, you will be looking for values of the equilibrium constants for the methane–steam reforming reaction:

$$CH_4 + H_2O \rightarrow CO + 3H_2 \qquad (A1.1)$$

And the water–gas shift reaction:

$$CO + H_2O \rightarrow CO_2 + H_2 \qquad (A1.2)$$

This appendix will simply highlight some of the points of understanding you should already have. As has been pointed out previously, this section is not a fundamental presentation of the thermodynamics being used but an attempt to present some of the relevant equations and data.

You should be aware of the existence of Gibbs free energy and the Gibbs free energy difference for a reaction expressed as ΔG. In your studies of thermodynamics applied to reactions by considering the state functions of the first and second laws, you will have been introduced to the Gibbs free energy and you will have realised that for a reaction at a stated temperature and pressure, it represents energy that is free to drive change, i.e. energy that will drive the reaction until *equilibrium* is reached.

Each reacting component has an associated Gibbs free energy of formation. At a given pressure, we will have a value of $(\Delta G_T)_f$. For a reaction of the type:

$$aA + bB \leftrightarrow cC + dD \qquad (A1.3)$$

we can generally write:

$$\Delta G_T = \left(\sum (n\Delta G_T)_f \right)_{products} - \left(\sum (n\Delta G_T)_f \right)_{reactants} \tag{A1.4}$$

where n is the number of moles of each component.
You will know, or realise, that if

$$\left(\sum (n\Delta G_T)_f \right)_{products} = \left(\sum (n\Delta G_T)_f \right)_{reactants} \tag{A1.5}$$

then ΔG becomes zero and effectively there is no driving force for change and we have a thermodynamic equilibrium.

The problem you and I have is that we want to quantify the amounts of each component at the thermodynamic equilibrium. This can be done by finding the n values in the last equation. This can be done, for example, by setting up a spreadsheet provided the component values of $(\Delta G_T)_f$ are known at the temperature and pressure of the reaction.

In the mass and energy balances represented here, we have adopted other thermodynamic equations to quantify our equilibrium amounts. Using a component value of $(\Delta G_T)_f$ in a mixture which may have been introduced to you as a chemical potential, an equation called the reaction isotherm has been produced:

$$\Delta G_T = \Delta G_T^\theta + RT \frac{a_C^c a_D^d}{a_A^a a_B^b} \tag{A1.6}$$

This can be written as a general equation but is reproduced here for the particular reaction already discussed:

$$aA + bB \leftrightarrow cC + dD \tag{A1.3}$$

The terms $a_A^a, a_B^b, a_C^c, a_D^d$, which can be written generally as a_i^n, are called component activities. Subscript i indicates the component, and superscript n is the appropriate component stoichiometric coefficient. We could write a book about activities, but let's not. For our simple, explanatory, appendix, we treat an activity as an appropriate thermodynamic concentration. That sounds good, but the ultimate aim of this appendix is to be able to quantify equilibrium amounts of components. So how do we use an activity?

Pure component and mixture fugacities can be defined and used. For our purposes, we assume that we have gained the knowledge previously or that we simply use the equations presented here.

We can write an activity a_i as:

$$a_i = \frac{f_i}{f_i^\theta} \tag{A1.7}$$

where a_i is the component activity, f_i is the component fugacity in the mixture and f_i^θ is the pure component fugacity at a standard state. Without exploring the further complexities of this area of thermodynamics, we will assume that for all our calculations, even at higher pressures, our gas mixtures behave ideally. If we take this assumption, then we can take a component fugacity and render it into a form that will enable us to achieve our goal of presenting a parameter that can quantify our component amounts at equilibrium.

For the assumption of gas mixture ideality, we can write:

$$a_i = \frac{p_i}{P^*} \tag{A1.8}$$

where p_i is a component partial pressure and P^* is a standard pressure.

You should know, or learn, that a partial pressure is the component pressure contribution to the total mixture pressure, P. Dalton's law simply says that $\sum p_i = P$. It is relatively simple to render this into a form that we use in our mass balancing in the main text:

$$p_i = y_i P \tag{A1.9}$$

where y_i is a gas-phase component mole fraction. If we are not comfortable with a mole fraction, then you are coming into this mass balancing too early. Go and revise your understanding of moles and the number of component moles divided by the sum of all the component moles in the mixture.

We are almost there in the quest for working equations that will give us the answers we want for equilibrium component amounts. Equation A1.6 gives us an equation for ΔG_T. We say that for a reaction, an equilibrium is reached when there is no driving force to produce change. In terms of the parameters we have considered, we want the total Gibbs free energy change for the reaction to be zero, i.e. $\Delta G_T = 0$. Thus, if we take Equation A1.6 and apply this, we will get:

$$\Delta G_T^\theta = -RT \frac{a_C^c a_D^d}{a_A^a a_B^b} \tag{A1.10}$$

Two important parameters appear in this equation. The first is the ratio:

$$\frac{a_C^c a_D^d}{a_A^a a_B^b}$$

In reaction thermodynamics, this ratio is called K, the equilibrium constant.

If the thermodynamic concentrations are written in terms of activities, then we consider K_a; if the activities are written as fugacities, then we consider K_f; if we consider the activities as partial pressures, as we will, then we consider K_p. We can then write the equilibrium reaction isotherm:

$$\Delta G_T^\theta = -RT \ln K_p \tag{A1.11}$$

Obviously, this equation depends on our being able to find values of ΔG_T^θ. This parameter requires a standard state to be fixed. We will refer it to a standard pressure of 1 bar and seek to subsequently calculate its numerical values at various values of temperature (T).

If we refer to Equation A1.4:

$$\Delta G = \left(\sum (n\Delta G_T)_f\right)_{products} - \left(\sum (n\Delta G_T)_f\right)_{reactants} \tag{A1.4}$$

then obviously this can be written as:

$$\Delta G_T^{\vartheta} = \left(\sum \left(n\Delta G_T^{\theta}\right)_f\right)_{products} - \left(\sum \left(n\Delta G_T^{\theta}\right)_f\right)_{reactants} \tag{A1.12}$$

Thus, at a stated temperature (often 298 K, 25°C) the values of ΔG_T^{ϑ} can be found and are available in published tables. Obviously, reactions usually take place at a variety of temperatures, and it is necessary to be able to adjust values of ΔG_T^{ϑ} to any temperature. The equation that offers this possibility is the Gibbs–Helmholtz equation:

$$\left\{\frac{\partial\left(\dfrac{\Delta G_T^{\theta}}{T}\right)}{\partial T}\right\}_P = \frac{-\Delta H_T^{\theta}}{T^2} \tag{A1.13}$$

We are dealing with as much reasonable simplification that we can get. On that basis, we find that for many reactions, over fairly wide temperature ranges, the variation in ΔH_T^{θ}, standard enthalpy change for the reaction at a temperature of T, is small (if enthalpy is not as memorable, understandable or clear as it should be, then there is a separate appendix that attempts to give some explanation). If we assume a constant pressure and ΔH_T^{θ} is almost constant, then we can carry out an integration to give a simplified form of the integrated Gibbs–Helmholtz equation:

$$\left[\frac{\Delta G_{T_2}^{\theta}}{T_2}\right] - \left[\frac{\Delta G_{T_1}^{\theta}}{T_1}\right] = \Delta H^{\theta}\left(\frac{1}{T_2} - \frac{1}{T_1}\right) \tag{A1.14}$$

It is often useful to combine this equation with Equation A1.11. This then gives:

$$\left[R\ln(K_{p_1})\right] - \left[R\ln(K_{p_2})\right] = \Delta H^{\theta}\left(\frac{1}{T_2} - \frac{1}{T_1}\right)$$

This can easily be arranged to give:

$$\ln\left[\frac{K_{p_2}}{K_{p_1}}\right] = -\frac{\Delta H^{\theta}}{R}\left(\frac{1}{T_2} - \frac{1}{T_1}\right) \tag{A1.15}$$

an equation sometimes known as the van't Hoff isotherm.

This gives a useful indication that the parameter $\ln(K_p)$ is a linear function of $\dfrac{1}{T}$.

This assumption will be made throughout the mass and energy balance equation carried out in this text. With the equations outlined in this appendix, it is possible to calculate values of ΔG_T^θ for a given reaction; calculate these at various temperatures using Equation A1.14; if necessary, express the ΔG_T^θ value as $\ln(K_p)$. The values of $\ln(K_p)$ at various temperatures are given below (Table A1.1):

A1.1 METHANE–STEAM REACTION

These results can be represented by an equation of the form:

$$\ln K_P = \frac{-27,106}{T} + 30.42$$

where T is in K.

A1.2 WATER–GAS SHIFT REACTION

These results can be represented by an equation of the form:

$$\ln K_P = \frac{4510.23}{T} - 4.196$$

where T is in K.

TABLE A1.1

$CO + H_2O \rightarrow CO_2$

K_p	$\ln(K_p)$	Temperature, T (K)
227.9	5.429	473
136.9	4.919	498
86.51	4.460	523
57.14	4.046	548
39.22	3.669	573
27.83	3.325	598
20.34	3.013	623
15.25	2.725	648
11.7	2.460	673
9.165	2.215	698
7.311	1.989	723
5.926	1.780	748
4.878	1.585	773
4.069	1.403	798
3.434	1.233	823
2.931	1.075	848
2.527	0.927	873
2.199	0.788	898
1.923	0.654	923

(*Continued*)

TABLE A1.1 (*Continued*)

$CO + H_2O \rightarrow CO_2$

K_p	$\ln(K_p)$	Temperature, T (K)
1.706	0.534	948
1.519	0.418	973
1.361	0.308	998
1.228	0.205	1023
1.113	0.107	1048
1.015	0.0149	1073
0.9295	−0.073	1098
0.8552	−0.156	1123
0.7901	−0.236	1148
0.7328	−0.311	1173
0.6822	−0.382	1198
0.6372	−0.045	1223
0.5971	−0.516	1248

$CH_4 + H_2O \rightarrow CO + 3H_2$

4.614×10^{-12}	−26.102	473
8.397×10^{-10}	−20.898	523
6.378×10^{-08}	−16.568	573
2.483×10^{-06}	−12.792	623
5.732×10^{-05}	−9.767	673
8.714×10^{-04}	−7.045	723
9.442×10^{-03}	−4.663	773
7.741×10^{-02}	−2.559	823
0.5029	−0.6870	873
1.189	0.173	898
2.686	0.988	923
5.821	1.761	948
12.14	2.497	973
24.42	3.195	998
47.53	3.861	1023
89.68	4.496	1048
164.4	5.102	1073
293.3	5.681	1098
510.1	6.235	1123
866.6	6.765	1148
1440	7.272	1173
2342	7.761	1198
3736	8.226	1223
5850	8.674	1248

Appendix 2
Calculation of Fractional Conversions for the Primary Reformer

To produce values of z_1 and z_2 as given in Table 1.3, a simple rearrangement was made. Basically, at the temperature in the primary reformer, 1093°K, the values of K_1 and K_2 for the two reactions considered were calculated as 275.80 and 0.91. Using the moles of each component in terms of z_1 and z_2 as given in Table 1.3, we can write:

For the methane–steam reaction:

$$K_1 = \frac{\left(\dfrac{98.6z_1 - 98.6z_1z_2}{463.3 + 2 \times 98.6z_1}\right) \cdot P \left(\dfrac{3 \times 98.6z_1 + 98.6z_1z_2}{463.3 + 2 \times 98.6z_1}\right)^3 \cdot P^3}{\left(\dfrac{(98.6 - 98.6z_1)}{463.3 + 2 \times 98.6z_1}\right) \cdot P \left(\dfrac{363.3 - 98.6z_1 - 98.6z_1z_2}{463.3 + 2 \times 98.6z_1}\right) \cdot P}$$

For the water–gas shift reaction:

$$K_2 = \frac{\left(\dfrac{98.6z_1z_2}{463.3 + 2 \times 98.6z_1}\right) \cdot P \left(\dfrac{3 \times 98.6z_1 + 98.6z_1z_2}{463.3 + 2 \times 98.6z_1}\right) \cdot P}{\left(\dfrac{(98.6 - 98.6z_1z_2)}{463.3 + 2 \times 98.6z_1}\right) \cdot P \left(\dfrac{363.3 - 98.6z_1 - 98.6z_1z_2}{463.3 + 2 \times 98.6z_1}\right) \cdot P}$$

Essentially, there are two equations and two unknowns. To obtain a single equation with two roots, we take the equation for K_2 and multiply it both sides by $\dfrac{K_1}{K_2}$. This gives us two equations for K_2. We simply subtract the equations from each other and then use SOLVER to find the values of z_1 and z_2 which minimise the difference. This then gives us the values of z_1 and z_2. SOLVER allows constraints to be specified for both z_1 and z_2; the obvious constraints were applied, and essentially, both values should lie between 0 and 1. The values were then found as quoted in the main text, 0.781 and 0.461.

Appendix 3

Discussion of Data Sources and Basic Equations Relating to the Enthalpy Property

In this appendix, various ways of obtaining the data required for the energy balance are laid out. To do the balance, the thermodynamic property of enthalpy is used. You should be familiar with this as a state property arising from the first law of thermodynamics. In terms of a system defined as a fixed mass of material, the flow of energy across the system boundary arises from the molecular energies associated with the material within the defined system. This energy is usually called the internal energy (U). The energy across the system boundary is transferred through two mechanisms: transfer as heat (q) and transfer as work (w). For the system defined, the change in internal energy can be made up of the energy transferred by the two mechanisms:

$$\Delta U = q + w$$

where ΔU is the change in internal energy associated with the transfer.

For changes of volume associated with chemical reaction at constant pressure, the energy transfer by a work mechanism can be written as:

$$w = -P\Delta V$$

where P is the pressure and ΔV is the change in volume produced by the chemical reaction. On this basis, we can write:

$$q = \Delta U + P\Delta V$$

All the terms on the right-hand side are state functions – i.e., they depend only on the initial and final states of the system – so we write the equation as:

$$q = \Delta H = \Delta U + P\Delta V$$

where ΔH is an enthalpy change.

At constant pressure, we can use this property to calculate energy transfers in the process for producing synthesis gas.

In finding values of enthalpy, this appendix explores various data sources that can be found and how to use the data sources to obtain useable enthalpy data. If you are consulting this appendix, you should already have some basic knowledge of the enthalpy property.

You should have already been introduced to the use of specific heats to find enthalpy values at various temperatures. Obviously, the necessary data have to be available. If these data are available, then the following approach can be used.

The necessary data are given in the table below:

TABLE A3.1
Polynomials for Selected Specific Heat Capacity at Constant Pressure as a Function of Temperature

Component	C_p (kmol/kmol K)
Hydrogen	$27.70 + 0.00081T$
Methane	$19.25 + 0.05213T + 1.197 \times 10^{-5}T^2 - 1.132 \times 10^{-8}T^3$
Nitrogen	$31.15 - 0.01357T + 2.680 \times 10^{-5}T^2 - 1.168 \times 10^{-8}T^3$
Ethane	$5.409 + 0.1781T - 6.938 \times 10^{-5}T^2 + 8.713 \times 10^{-9}T^3$
Propane	$-4.224 + 0.3063T - 1.586 \times 10^{-4}T^2 + 3.215 \times 10^{-8}T^3$
Butane	$9.487 + 0.3313T - 1.108 \times 10^{-4}T^2 - 2.822 \times 10^{-9}T^3$
Water	$32.24 + 1.924 \times 10^{-3}T + 1.055 \times 10^{-5}T^2 - 3.596 \times 10^{-9}T^3$
Hydrogen sulphide	$31.94 + 0.001436T + 2.432 \times 10^{-5}T^2 - 1.176 \times 10^{-8}T^3$

Values for the constants in the C_p polynomials are taken from the following reference:
Robert C Reed, John M Prausnitz, Bruce E Poling, Fourth Edition 1987, *The Properties of Gases and Liquids*, McGraw-Hill Book Company.

The specific heat capacities are given in polynomial form as a function of temperature (T).

Determining a component enthalpy and subsequently a stream enthalpy depends on the use of the defining equation:

$$\Delta H = \int_{T_1}^{T_2} mC_p \, dT$$

where m is a component mass, C_p is the specific enthalpy at constant pressure and T is a temperature in appropriate thermodynamic temperature units. The component enthalpy can be calculated by defining a reference temperature T_1 and an enthalpy at that temperature. It is then possible to calculate the component enthalpy at any temperature T_2.

Often the C_p is expressed as a polynomial expression in T. In this case, the C_p is expressed as:

$$C_p = a + bT + cT^2 + dT^3$$

where a, b, c and d are constants in the polynomial expression.

The values of constants for various components in the production of the ammonia syngas are given in Table A3.1 where the appropriate polynomials are reproduced.

For a component, the expression can be written as:

$$\Delta H = \int_{T_1}^{T_2} m\left(a + bT + cT^2 + dT^3\right)dT$$

Assuming integration between limits comes easily to mind, we can write:

$$\Delta H = m\left[\left(aT + \frac{bT^2}{2} + \frac{cT^3}{3} + \frac{dT^4}{4}\right)\right]_{T_1}^{T_2}$$

If the mixture (stream enthalpy) is required, then the value can be found:

$$\Delta H_{\text{stream}} = \int_{T_1}^{T_2} \sum_{1}^{N} \left(m_i C_P dT\right)$$

where N is the total number of components. The individual component enthalpies can be calculated through the integrated expression presented and then summed as indicated.

In assigning values of enthalpy, if calculations of the type just indicated using C_P values are not possible or values not available, it is useful to find values of the enthalpy from other sources. In most of the calculations in this text, the enthalpies are calculated through appropriate polynomials produced from known enthalpy data.

Because we have equations for ΔH, we normally carefully write the difference down as:

$$\Delta H = H_2 - H_1$$

We often define H_1 as a reference state at which we fix a value of H_1, and knowing the value of H_1, we can assign a value to H_2. In the enthalpy data used in the calculations outlined in this text, various reference states are used.

The first set of data given below tabulates the values of component enthalpies for each of the reacting species at various temperatures. The values of enthalpy vary with temperature, and it is necessary to have component enthalpy values at different temperatures to be able to calculate the necessary energy balances. As already stated, various sources of enthalpy data are used, and the next sections briefly give the genesis of each.

A3.1   VALUES OF ENTHALPY AT VARIOUS TEMPERATURES ASSUMING ALL COMPONENTS TO BE IDEAL GASES WITH THE TEMPERATURE AT REFERENCE STATE 1 TAKEN AS 0°K.

Data in this section are presented where the enthalpy at 0°K is set to zero and all components are considered as ideal gases. The enthalpies can be calculated at temperatures above 0°K. It is not proposed to give a detailed outline of the calculations other than to note values reported in Table A3.2 below, and these values are based on the equation that is used to calculate enthalpy.

The working equation for these data is:

$$\Delta H = \Delta H_F^0 + (H_T^o - H_0^o)$$

where ΔH_F^0 is the enthalpy of formation for a component at the appropriate reference temperature (in this case, $0°K$), i.e. the enthalpy change when the component is formed from its constituent elements at the designated temperature; H_T^o is a standard enthalpy at any temperature T; and H_0^o is an enthalpy related to a temperature of $0°K$.

The values reported are based on the assumption that at $0°K$, H_0^o is zero. The numerical values reported are values of $\Delta H = \Delta H_F^0 + H_T^o$, where H_T^o is an enthalpy related to temperature T.

TABLE A3.2
Values of $\Delta H = \Delta H_F^o + H_T^o$, where H_T^o is an Enthalpy Related to Temperature T

Temperature (°K)	Methane	Hydrogen	Nitrogen
298	−5.6719E+04	8.6622E+03	8.6456E+03
400	−5.2995E+04	1.1612E+04	9.1033E+03
500	−4.8663E+04	1.4562E+04	1.4819E+04
600	−4.3704E+04	1.7569E+04	1.7875E+04
800	−3.2094E+04	2.3756E+04	2.3952E+04
1000	−1.8540E+04	3.0172E+04	2.9943E+04
1500	2.1372E+04	4.7214E+04	4.4213E+04

Temperature (°K)	Oxygen	Water	Carbon Monoxide	Carbon Dioxide
298	8.5795E+03	−2.2920E+05	−1.0510E+05	−3.8371E+05
400	1.1658E+04	−2.2583E+05	−1.0467E+05	−3.8322E+05
500	1.4757E+04	−2.2246E+05	−9.9139E+04	−3.7608E+05
600	1.7934E+04	−2.1903E+05	−9.6015E+04	−3.6800E+05
800	2.4501E+04	−2.1198E+05	−8.9409E+04	−3.6140E+05
1000	3.1324E+04	−2.0475E+05	−8.2298E+04	−3.5030E+05
1500	4.9270E+04	−1.8604E+05	−6.2119E+04	−3.2237E+05

Units of enthalpy are kJ/kmol.

In this case, these data were then fitted using a polynomial of the form:

$$\text{Enthalpy}(H) = a + bT + cT^2 + dT^3$$

The polynomial coefficients a, b, c and d have been fitted for each component using SOLVER in an Excel spreadsheet, and the values are set out in Table A3.3.

TABLE A3.3
The Polynomial Coefficients a, b, c and d Fitted for Each Component Using the Data in Table A3.2

Component	a	b	c	d
Methane	−63,110	9.291	4.31E − 02	−7.85E − 06
Hydrogen	385.5	26.92	4.31E − 02	0
Nitrogen	−373.1	29.93	1.43E − 03	1.05E − 06
Water	−2.39E+05	30.31	4.33E − 03	7.85E − 07
Carbon monoxide	−113,000	25.348	4.844E − 03	6.105E − 07
Carbon dioxide	−3.893E+05	10.047	0.04255	−1.275E − 05
Oxygen	114.84	27.0	0.00493	−7.246E − 07

The enthalpy calculated has units of kJ/kmol. These data and polynomials have been used in carrying out the energy balance for most of the operations described in this text. The data were used by Hampson in his original calculations.

The data were fitted to the polynomial by inputting guessed constants into the appropriate polynomial and instructing SOLVER to reduce the difference between the enthalpy values then calculated and the actually tabulated value to zero.

A3.2 VALUES OF ENTHALPY AT VARIOUS TEMPERATURES ASSUMING ALL COMPONENTS TO BE IDEAL GASES WITH THE FORMATION COMPONENTS CONSIDERED AS IDEAL GASES AT 0°K

The data values quoted in this section are taken from Refs. [1] and [2].

The units of the quoted enthalpy are kJ/mol. The elements within the table (hydrogen, nitrogen and oxygen) have comparable values to those quoted in Table A3.2 because the formation enthalpies of elements are set at zero. The values tabled for methane, water, carbon monoxide and carbon dioxide are, as indicated in the original table, values of $(H_T^o - H_0^o)$. To make them comparable to values in Table A3.2, they require values of ΔH_F^0, i.e. the values of the enthalpy of formation at 0°K for the methane, water, carbon monoxide and carbon dioxide. These values are available [3].

The values are as follows:

Component	ΔH_F^0 at 0°K kJ/mol
Methane	−66.564
Water	−238.931
Carbon dioxide	−393.108
Carbon monoxide	−113.803

If these values are appropriately added to the figures in Table A3.4, then we obtain the following:

TABLE A3.4
Values of Enthalpy at Various Temperatures Assuming all Components to be Ideal Gases with the Formation Components Considered as Ideal Gases at 0°K

T (°C)	Methane	Hydrogen	Water	Carbon Monoxide	Carbon Dioxide	Nitrogen	Oxygen
298	10.029	8.468	9.906	8.672	9.364	8.6705	8.661
400	13.903	11.426	13.364	11.647	13.367	11.6415	11.681
500	18.263	14.349	16.843	14.602	17.669	14.581	14.744
600	23.217	17.278	20.427	17.613	22.270	17.564	17.903
800	34.815	23.168	27.989	23.849	32.173	23.717	24.501
1000	48.367	29.145	36.016	30.363	42.769	30.135	31.367
1500	88.408	44.745	57.940	47.522	71.145	47.085	49.270

TABLE A3.5
Values of Enthalpy Calculated Using Values in Table A3.4 Combined with Values of ΔH_F^0 at 0°K Quoted in the Text

T K	Methane	Hydrogen	Water	Carbon Monoxide	Carbon Dioxide	Nitrogen	Oxygen
298	−56.535	8.468	−229.025	−105.214	−383.744	8.6705	8.661
400	−52.661	11.426	−225.567	−102.156	−379.741	11.6415	11.681
500	−48.301	14.349	−222.088	−99.201	−375.439	14.581	14.744
600	−43.347	17.278	−218.504	−96.190	−370.838	17.564	17.903
800	−31.749	23.168	−210.942	−89.954	−360.935	23.717	24.501
1000	−18.917	29.145	−202.915	−83.440	−350.339	30.135	31.367
1500	21.844	44.745	−180.991	−66.281	−322.369	47.085	49.270

Given the diverse sources of data, the agreement between the figures given in Table A3.2 and Table A3.5 is considered acceptable.

These values can also be fitted to a polynomial expression of similar form to that set out earlier in this appendix, namely:

$$\text{Enthalpy}\,(H) = a + bT + cT^2 + dT^3$$

For this polynomial, the constants a, b, c and d are as follows (Table A3.6):

TABLE A3.6
Enthalpy Values From Table A3.5 Fitted to a Polynomial of the Form:
Enthalpy $(H) = a + bT + cT^2 + dT^3$

Component	a	b	c	d
Methane	−63,496.1	12.194	0.03971	−6.6E−06
Hydrogen	−221.507	29.3	−7.0E−05	7.67E−07
Nitrogen	114.34	28.387	−7.0E−04	7.93E−07
Water	−2.42435E + 05	34.887	−8.5E−04	2.41E−06
Carbon monoxide	−1.1343E + 05	26.495	3.682E−03	−2.80E−07
Carbon dioxide	−3.9015E + 05	1.047E+01	4.255E−02	−1.301E−05
Oxygen	114.844	27.0	4.93E-03	−7.2E−07

If the values of enthalpy found using these constants are then used in the primary reformer energy balance, then the following table (Table A3.7) can be drawn up:

TABLE A3.7
Values of Enthalpy Found Using the Constants in Table A3.6 Used in the Primary Reformer Energy Balance.

Enthalpy/mol	Input Enthalpy	Enthalpy/mol	Output Enthalpy
−3.975E+04	−3.764E+06	−9.411E+03	−2.030E+05
−4.448E+04	−1.335E+05	6.941E+03	
−4.833E+04	−2.416E+04	2.635E + 04	
−5.200E+04	−2.080E+04	4.466E + 04	
1.914E + 04	2.680E+04	3.223E+04	4.513E+04
−2.186E+05	−7.942E+07	−2.009E+05	−5.038E+07
−9.402E+04	0.000E+00	−7.943E+04	−3.297E+06
1.970E+04	0.000E+00	3.368E+04	8.979E+06
−3.681E+05	0.000E+00	−3.436E+05	−1.221E+07
Total input	−8.334E+07	Total output	−5.706E+07
	Output–input=	2.62735E + 07	

This value, 2.62735E+07, is directly comparable to the value calculated in the primary reformer section.

A3.3 VALUES OF ENTHALPY AT VARIOUS TEMPERATURES ASSUMING ALL COMPONENTS TO BE IDEAL GASES USING A VAN'T HOFF BOX AND MEAN HEAT CAPACITIES

Elements of the van't Hoff box can be used to carry out a typical first-law calculation. The van't Hoff box for this situation exists as follows:

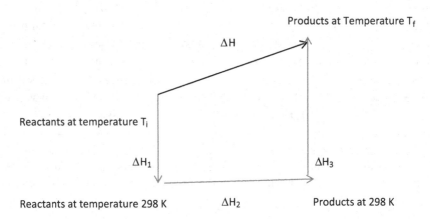

As indicated in the previous calculations, the enthalpy change going from state i to state f including the change due to reaction can be calculated directly from the enthalpy at T_i and the enthalpy at T_f calculated using the appropriate polynomials.

Thus, ΔH can be calculated as:

$$\Delta H = \sum \left(\text{Product Enthalpies at } T_f \right) - \sum \left(\text{Reactant Enthalpies at } T_i \right)$$

For comparison, using the elements of the van't Hoff box, the enthalpy ΔH could be calculated as:

$$\Delta H = \Delta H_1 + \Delta H_2 + \Delta H_3$$

where:

$$\Delta H_1 = \sum \left(MC_p (298 - T_i) \right)_{\text{Reactants}}$$

$$\Delta H_2 = \sum \left(\Delta H_f^0 \right)_{\text{Products}} - \sum \left(\Delta H_f^0 \right)_{\text{Reactants}} \quad \text{at } 298\,K$$

$$\Delta H_3 = \sum \left(MC_p (T_f - 298) \right)_{\text{Products}}$$

M represents component moles, C_p is a specific heat capacity at constant pressure, T_f and T_i are final and initial temperatures and ΔH_f^0 is the enthalpy of formation of a component.

If we consider the calculation using the van't Hoff box, the use of mean molar heat capacities $\widetilde{C_p}$ is also considered. This is defined as:

$$\widetilde{C_p} = \frac{\displaystyle\int_{298}^{T_f} C_p(T)\,dT}{\displaystyle\int_{298}^{T_f} dT} = \frac{\displaystyle\int_{298}^{T_f} C_p(T)\,dT}{\left(T_f - 298\right)}$$

Thus, a constant value of $\widetilde{C_p}$ can be written at a particular value of T_f. Using the tables of values given in Ref. [4], the following table can be constructed:

Using values of mean specific heat capacities the following table can be constructed for the van't Hoff box:

Using these figures, it is possible to calculate ΔH_1, ΔH_2 and ΔH_3:

$$\Delta H_1 = M\overline{C_p}(298 - T_i)(\text{Input})$$

$$\Delta H_2 = \sum \left(M\Delta H_f\right)_{\text{Products}} - \sum \left(M\Delta H_f\right)_{\text{Reactants}}$$

$$\Delta H_3 = M\overline{C_p}\left(T_f - 298\right)(\text{Output})$$

Using the figures in Table A3.8 (see page 116):

$$\Delta H_1 = 17{,}531.082(298 - 673) = -6{,}574{,}156 \text{ kJ}$$

$$\Delta H_2 = -80{,}818{,}599 - (-95{,}297{,}717) = 14{,}479{,}119 \text{ kJ}$$

$$\Delta H_3 = 21{,}689.84(1123 - 298) = 17{,}894{,}121 \text{ kJ}$$

Thus, $\Delta H_1 + \Delta H_2 + \Delta H_3 = -6{,}574{,}156 + 14{,}479{,}119 + 17{,}894{,}121 = 25{,}799{,}084 \text{ kJ}$

This value can be compared with the value calculated in the section for the primary reformer, 2.627E+07.

The values calculated in different ways from different data sources agree within less than 2%.

A3.4 VALUES OF ENTHALPY AT VARIOUS TEMPERATURES USING POLYNOMIALS PUBLISHED BY NASA

As a final check, thermodynamic data published by NASA [5] were used.

The data are presented in polynomial form, and a polynomial value for enthalpy is presented as:

$$\frac{H}{RT} = a1 + a2\frac{T}{2} + a3\frac{T^2}{3} + a4\frac{T^3}{4} + a5\frac{T^4}{5} + \frac{a6}{T}$$

TABLE A3.8
Using Values of Mean Specific Heat Capacities the Following Table Can Be Constructed for the Van't Hoff Box:

Van't Hoff Box	Input T_i = 673 K					Output T_f = 1123 K				
	M	ΔH_F	$M\Delta H_F$	C_p	MC_p	M	ΔH_F	$M\Delta H_F$	C_p	MC_p
CH_4	94.68	−74847	−708,6514	45.92	4347.71	21.57	−74847	−161,4765	57.58	1242
C_2H_6	3	−84667	−254,001	75.72	227.16	0	−84667	0		
C_3H_8	0.5	−103847	−51,923.5	109.12	54.56	0	−103847	0		
C_4H_{10}	0.4	−124733	−49,893.2	143.46	57.38	0	−124733	0		
N_2	1.4	0	0	29.7	41.58	1.4	0	0	31.0	43.4
H_2O	363.3	−241862	−87,855,386	35.24	12,802.69	250.75	−241862	−60,638,241	37.82	9483.4
CO	0	−110520	0		0	41.50	−110520	−458,6925	31.32	1300
H_2	0	0	0		0	266.60	0	0	29.6	7891.4
CO_2	0	−393514	0		0	35.52	−393514	−139,78668	48.74	1731.2
			Sum $M\Delta H_F$ −95,297,717		Sum MC_p 17,531.08			Sum $M\Delta H_F$ −80,818,599		Sum MC_p 21,689.80

where $a1$–$a6$ are polynomial constants, T is the temperature in °K and H is the enthalpy defined as:

$$H = \Delta H_F(298) + \big(H(T) - H(298)\big)$$

The value of H is calculated relative to a temperature of 298°K. Given the form of the NASA polynomial, it would be extremely difficult to adjust the figures to a comparable base for the polynomials previously reported. In the event, because, as usual, the energy balance is dealing in enthalpy differences, the answers obtained using the NASA data should be comparable.

The input temperature to the primary reformer has been set at 673°K and the exit temperature at 1123°K. The NASA polynomial gives the values indicated in Table A3.9. For the purpose of this calculation, the higher hydrocarbons were included in the figure for methane.

TABLE A3.9

Energy Balance on the Primary Reformer Using Data Calculated Using the NASA Polynomial

Enthalpy/mol	Input Enthalpy	Enthalpy/mol	Output Enthalpy
−5.741E+04	−5.659E+06	−2.647E+04	−5.710E+05
−2.286E+05		6.941E+03	
1.096E+04		2.635E+04	
−5.200E+04		4.466E+04	
1.112E+04	1.556E+04	2.553E+04	3.574E+04
−2.286E+05	−8.306E+07	−2.106E+05	−5.282E+07
−9.928E+04	0.000E+00	−8.446E+04	−3.505E+06
1.096E+04	0.000E+00	2.437E+04	6.498E+06
−3.771E+05	0.000E+00	−3.533E+05	−1.255E+07
Total input	−8.871E+07	Total output	−6.291E+07
	Output–input	2.580E+07	

If we review the four data sources outlined in this appendix and their application to the energy balance on the primary reformer, then we can review the figures calculated as necessary energy to be input to the primary reformer:

Source Number	Energy Input Required (kJ/kmol)
1	2.6273E+07
2	2.580E+07
3	2.5799E+07
4	2.5709E+07

This gives a maximum variation of just over 2%.

This appendix has explored possible sources of enthalpy data for energy balances. It has laid out some theoretical approaches to obtaining enthalpy data. The enthalpy polynomials based on the data set out in Table A3.2 have been used as the source of enthalpy data in the energy balances.

REFERENCES

1. API Research Project 44, Carnegie Institute of Technology, 1953, *Selected Values of Physical and Thermodynamic Properties of Hydrocarbons and Related Compounds*, White Papers: University of Nebraska-Lincoln Libraries, Pittsburgh.
2. The values are reproduced by A M Mearns, 1973, *Chemical Engineering Process Analysis*, Oliver and Boyd, Edinburgh, UK, p238.
3. B Ruscic, 2015, Active Thermochemical Tables (ATcT) values based on ver. 1.118 of the Thermochemical Network; available at ATcT.anl.gov).
4. Edward V Thompson, William H Ceckler, 1978, *Introduction to Chemical Engineering*, McGraw-Hill International Edition, New York, USA.
5. W C Gardiner, Ed, 1984, *Thermochemical Data for Combustion Calculations*, Chapter 8 of Combustion Chemistry, Springer-Verlag, New York, USA.

Appendix 4
Solution of Simultaneous Equations for the Secondary Reformer

As mentioned in the main text, it is possible to solve the five equations simultaneously using the SOLVER function in an Excel spreadsheet. In this particular problem, there were five components:

Carbon monoxide (M)
Carbon dioxide (D)
Nitrogen (N)
Hydrogen (H)
Water (steam) (W)

The equations to be solved have been described in the main text; they were produced from C, H and O atom balances and the N/H stoichiometric ratio requirement, and a fifth equation was supplied by the thermodynamic equilibrium relationship through the value of the equilibrium constant for the water–gas shift reaction.

All five equations were rearranged to equal zero. Initial values had to be supplied for the moles of the five components. For the carbon dioxide, carbon monoxide, hydrogen and water, the values supplied were the initial moles entering the reactor; for the nitrogen, a value of the entering hydrogen/3 was supplied. For each equation, a value was calculated based on the supplied initial values. The values should, obviously, be zero for a theoretical solution. To obtain a solution, each value of the equation solutions obtained was summed and SOLVER was instructed to minimise the absolute value of this sum by changing the values of the number of moles of each component in the balance equations. SOLVER was instructed that none of the component moles should go negative; the GRG non-linear engine was used and the following solution (Table A4.1) obtained:

TABLE A4.1

Comparison between Solver and Calculated Solutions

Component	SOLVER Solution (kmol)	Calculated Solution (Based on M = 64.41 kmol)
CO	64.41	64.41
CO_2	33.40	33.2
N_2	111.00	111.19
H_2	268.59	268.59
H_2O	291.05	291.05

This was considered to be acceptable, and the calculated values for the secondary reformer output were taken as the output moles.

To emphasise the usefulness of the procedure described, the necessary manipulation of the simultaneous equations is shown below:

$$D = 97.81 - M = 97.61 - 64.41 = 33.2 \text{ kmol}$$

We know:

$$363.21 + 0.532N = M + 2D + W$$

Thus:

$$363.21 + 0.532N = 64.41 + 2 \times 33.2 + W$$

And then:

$$W = 231.3 + 0.532N$$

We know:

$$1119.24 = 2H + 2W$$

Hence:

$$W = \frac{1119.24 - 2H}{2} = 559.62 - H$$

Hence:

$$559.62 - H = 231.3 + 0.532N$$

$$H = 328.32 - 0.532N$$

We know:

$$M + H = 3N$$

Thus:

$$64.41 + 328.32 - 0.532N = 3N$$

$$N = 111.19 \text{ kmol}$$

Thus:

$$M + H = 3N$$

And hence:

$$64.41 + H = 3 \times 111.19$$

$$H = 269.17 \text{ kmol}$$

And:

$$W = \frac{1119.24 - 2H}{2} = 559.62 - H$$

Thus:

$$W = 559.62 - 269.17 = 290.45 \text{ kmol}$$

If we have a calculation of the amount of N_2 introduced into the reformer, then we can calculate the entering oxygen in the air:

The amount of nitrogen was calculated as 111.19 kmol; if we take (as we have) the ratio of O_2/N_2 in the air as 21/79, then the mount of O_2 introduced is $111.19 \times 21/79 = 29.56$ kmol.

In summary:

Component	Output kmol from the Secondary Reformer
Carbon monoxide	64.41
Carbon dioxide	33.2
Steam	290.45
Nitrogen	111.19
Hydrogen	269.17

Appendix 5
Use of Approach to Equilibrium in Assessing Catalyst Performance

Catalysts cannot change the equilibrium outlet compositions for given reactions in reactors. They can influence the speed at which equilibrium is approached, but the ultimate equilibrium conditions at a given temperature and pressure are determined by thermodynamic considerations.

It is sometimes convenient to have an easy measure of how effective a given catalyst is in reaching a desired outlet composition. If the outlet temperature and pressure of a process stream are known, then thermodynamically, an equilibrium constant can be calculated and the theoretical equilibrium composition fixed; the actual output compositions can also be measured. These compositions can then be used to calculate an 'operating equilibrium constant' based purely on these measured outlet compositions. The compositions are substituted into the appropriate equilibrium constant equation.

This 'operating equilibrium constant' can then be used to calculate the temperature corresponding to its value using an appropriate expression or graph giving equilibrium constant as a function of temperature.

As an example, if we take the water–gas shift reaction

$$CO + H_2O \leftrightarrow CO_2 + H_2$$

And the K_p for the reaction can be calculated from:

$$\ln K_p = \frac{4510.23}{T} - 4.196$$

We know we can write an equation for K_p as:

$$K_P = \frac{p_{CO_2} p_{H_2}}{p_{CO} p_{H_2O}}$$

The actual reactor outlet compositions have been measured as follows:

This gives a value of K, 'the operating equilibrium constant', at the operating temperature of 900 K as:

$$K_P = \frac{0.167 \times 0.39}{0.084 \times 0.359} = 2.16$$

TABLE A5.1

The Actual Reactor Outlet Compositions for the Water Gas Shift Reaction at 867.7 K

Component	kmol	Mole Fraction (y)
Carbon dioxide	70.0	0.167
Hydrogen	163.0	0.39
Carbon monoxide	35.0	0.084
Water	150.0	0.359

If we now use the equation for $\ln K_P$ as a function of $1/T$, we can find a temperature corresponding to this value of K_P. We find $T = 882.3$ K.

We know the output temperature is 900 K, and the reactor has produced a product stream that corresponds to a value of K_P that would be calculated at a temperature of 882.3 K. Thus, we measure the 'effectiveness' of the catalyst as getting to equilibrium in terms of the difference in temperature between the actual equilibrium temperature and the temperature corresponding to the actual conversion achieved. In this illustrative case, the difference is $900 - 882.3 = 17.7$ K. This difference is called a measure of approach to equilibrium and measures the effectiveness of the catalyst in getting the reacting mixture to equilibrium.

In this case, we have an exothermic reaction: the higher the temperature, the smaller the equilibrium constant. If we had an operating temperature of, say, 850 K, then using the approach to equilibrium as a measure of the effectiveness of the catalyst, we would use an equilibrium constant at a value of $850 + 17.7 = 867.7$ K for mass balance calculations.

Index

Printed in the United States
by Baker & Taylor Publisher Services